Ignite Student Intellect and Imagination in Language Arts

Sandra L. Schurr and Kathy L. LaMorte

National Middle School Association
Westerville, Ohio

Sue Swaim, Executive Director
Jeff Ward, Deputy Executive Director
April Tibbles, Director of Publications
Edward Brazee, Editor, Professional Publications
John Lounsbury, Consulting Editor, Professional Publications
Mary Mitchell, Designer, Editorial Assistant
Dawn Williams, Publications Manager
Marcia Meade-Hurst, Senior Publications Representative
Nikia Reveal, Graphic Designer
Lindsay Kronmiller, Graphic Designer

Library of Congress Cataloging-in-Publication Data
Schurr, Sandra
Ignite student intellect and imagination in language arts/Sandra L. Schurr and Kathy L. LaMorte,
p. cm
ISBN 978-1-56090-203-4
1. Language arts (Middle school)—United States. 2. Language arts (Middle school)—Activity programs. I. LaMorte, Kathy. II. Title
LB1631.S284 2007

National Middle School Association
4151 Executive Parkway, Suite 300
Westerville, OH 43081
1-800-528-NMSA
NMSA®

Contents

Introduction: Discover Bloom's Taxonomy

Discover Bloom's Taxonomy—and how its use will ignite students' intellect and imagination in mastering the skills and concepts called for by the standards of the National Council of Teachers of English and International Reading Association.

Bloom's Taxonomy has been used by educators worldwide and provides the organizing structure for almost all the rich and practical resources in this book. Developed many years ago, by Benjamin Bloom and his colleagues at the University of Chicago, this cognitive taxonomy defines a stairway of six levels of learning ranging from the lowest level—*knowledge*—up to the highest level—*evaluation.*

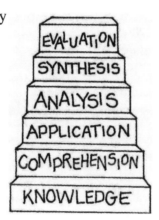

Teachers have found this taxonomy easy to understand and most useful as a basis for planning instruction and activities that will ensure learning goes beyond the mere acquisition of information. The high stakes testing now going on throughout the United States is directly correlated with subject-area national standards that use Bloom's Taxonomy as a primary model for designing specific items in middle school reading, mathematics, science, and social studies tests. In fact 50% of the test items for sixth graders, 60% of test items for seventh graders, and 70% of test items for eighth graders focus on Bloom's four higher order thinking skill levels: *application, analysis, synthesis, evaluation.* Using Bloom's Taxonomy as an organizing structure while teaching skills and content in all of the major subject areas is, then, a sound way to help ensure student success.

On pages 3-4 the Six Levels of Learning in Bloom's Taxonomy are defined together with a list of the most common verbs or behaviors assigned to each level. These two pages should be reproduced and copies given to all students to place in their notebooks as ready reminders about the types of tasks and thinking that are required as one moves from one level of learning to the next. In addition, an enlarged copy of these pages might well be made into a poster for permanent display in the classroom where it will serve as a visible reminder for both students and teachers of the need to always take learning to the higher levels.

Part I—Bloom Sheets

Following the format of Bloom's Taxonomy, we created multi-use Bloom Sheets, single pages that each contain an activity to be completed at each level of learning and that address specific standards of the National Council of Teachers of English (NCTE) and International Reading Association (IRA). The tasks require a lot of thinking and investigations that involve other subject areas and skills. The variety of topics touched on in these sheets and the several processes employed make them valuable in achieving a good, general education.

1

The Bloom Sheets are very popular with teachers because they

- Include both lower and higher levels of thinking skills.
- Present the information and assigned tasks in a straightforward manner.
- Use the same key verbs and behaviors that are employed in constructing both state and national tests.
- Are a built-in tool for differentiating instruction.

The layout and design of the Bloom Sheets are the result of considerable experimentation with middle school students. The typeface selected and the hand-created graphics by Kathy LaMorte have been demonstrated to immediately attract students and engage them in a positive way.

Classroom teachers by the hundreds have been delighted with the Bloom Sheets and their multiple uses. They can be the basis for independent study assignments, classroom work by individuals or small groups, or as homework. They can also be used as the organizing structure for learning stations, a multidisciplinary unit, a research paper, or as special assignments or enrichment tasks. Still additional uses of the taxonomy are suggested in Part II and Part III.

Part I contains 93 Bloom Sheets organized in groups that relate to the 12 *Standards for English Language Arts* proposed by NCTE and IRA.

Part II—Imaginative Assessment Options and Real-Life Applications

This section is chock-full of interesting ideas that will motivate students and lead to their mastering various concepts. First, there are four lists of different ways to assess learning in language arts. These lists offer ideas for including artifacts in students' portfolios, for creating language arts-related products, and for planning and giving language arts performances. There are enough possibilities here to whet students' imagination and get them actively involved in demonstrating their learning in ways other than by a paper and pencil test.

Following these assessment options are nine sheets that offer action-oriented investigations in ways language arts applies to real life. These could be completed by individuals or small groups, but either way they will bring relevance to language arts.

Part III—More Instructional Tools and Techniques

The last section contains a potpourri of suggestions for ways to enhance the teaching of language arts in the middle level classroom. It starts off with "Ten Smart Ways to Use Bloom's Taxonomy in the Language Arts Classroom," which is in itself a gold mine of great ideas. Then 10 creative ways to write reports are offered followed by a number of other lists and guidelines. These imaginative tools and techniques will bring new life and vigor to language arts classes.

Take time in the beginning to get acquainted with the abundant and rich resources contained in these pages, for you will discover a wealth of engaging ideas that you will turn to time after time as you help youngsters understand and enjoy the world of language arts. The book's spiral binding makes the Bloom Sheets and other reproducibles easy to copy.

Bloom's Taxonomy

Essential Levels of Learning	Most Common Verbs or Behaviors at this Level
1. Knowledge This level involves the basic recall of data or information and is the foundation for the other levels that follow. Although knowledge provides a basis for higher levels of thinking, it is simply a starting point. It is important for learners to internalize the information but recognize this level only involves recalling or restating the information.	*Choose, count, define, distinguish, draw, find, know, label, list, locate, match, memorize, name, pick, point, read, recall, recite, recognize, record, reproduce, select, state, trace, underline*
2. Comprehension This level focuses on one's ability to grasp meaning and to understand the basic information presented by translating, interpreting, and extrapolating it. At this level it is important for learners to demonstrate their ability to use the information in some way even when the teacher provides them with cues or signals.	*Associate, classify, conclude, demonstrate, describe, determine, differentiate, expand, explain, extend, find, generalize, give examples, give in own words, illustrate, interpret, measure, paraphrase, prepare, reorder, recognize, retell, reword, rewrite, restate, show, suggest, summarize, translate*
3. Application This level requires the student to make use of learned material in new situations. It also mandates that the learner be able to apply the information, ideas, or skill sets in more than one academic setting. At this level it is important for learners to use the information in a context different from the one in which it was taught without any cues or help from the teacher.	*Apply, calculate, collect information, complete, compute, construct, convert, derive, demonstrate, develop, discover, discuss, employ, examine, experiment, find, graph, interview, investigate, locate, make, model, organize, perform, plan, prepare, present, produce, prove, record, relate, show, solve, use*

4. Analysis This level involves the separating of items, materials, or ideas into their component parts and showing the relationship between and among those parts. At this level it is important for learners to acquire an understanding of both the content and the structural form of the material.	*Analyze, categorize, compare, contrast, debate, deduce, diagram, discover, divide, draw conclusions, examine, form generalizations, group, infer, outline, point out, relate, sort, subdivide, survey, take apart, uncover.*
5. Synthesis This level is the opposite of the Analysis Level because it involves the ability to put together separate ideas to form a new and different whole or umbrella idea or to establish new relationships between or among its parts. At this level it is important for learners to put their ideas and knowledge together in new and different ways leading to innovation and creativity.	*Arrange, build, combine, compose, create, derive, design, devise, develop, formulate, generate, imagine, integrate, invent, make up, originate, organize, perform, plan, prepare, present, produce, propose, rearrange, revise, rewrite, synthesize*
6. Evaluation This level tests one's ability to judge the worth or value of material and ideas against stated criteria and then defend the result. Remember that at this level it is important for learners to review and assert evidence, facts, ideas, and alternatives for solving problems and making decisions that can be supported by appropriate criteria and thoughtful justifications.	*Appraise, argue, assess, award, choose, consider, critique, defend, discriminate, evaluate, grade, judge, justify, measure, rank, rate, recommend, support, test, validate, verify*

Ignite Student Intellect and Imagination in Language Arts, by Sandra L. Schurr and Kathy L. LaMorte. Published by National Middle School Association, 2007.

PART I
Bloom Sheets

The Bloom Sheets that follow are organized in sections. Each section includes a number of Bloom Sheets that address primarily one of the 12 standards proposed by National Council of Teachers of English (NCTE) and International Reading Association (IRA). These standards are recognized as the national standard for language arts and are used in planning programs and assessment measures by leading publishers and curriculum designers.

Standards for the English Language Arts
Excerpt of Chapter 3

The standards presented define what we believe students should know and be able to do in the English language arts. We believe that these standards should articulate a consensus growing out of actual classroom practices and not be a prescriptive framework. If the standards work, then teachers will recognize their students, themselves, their goals, and their daily endeavors in this document; so, too, will they be inspired, motivated, and provoked to reevaluate some of what they do in class. By engaging with these standards, teachers will, we hope, also think and talk energetically about the assumptions that underlie their own classroom practices and those of their colleagues.

The standards reflect some of the best ideas already at work in English language arts education around the country. Because language and the language arts continue to evolve and grow, our standards must remain provisional enough to leave room for future developments in the field. And it is important to reemphasize that these standards are meant to be suggestive, not exhaustive. Ideally, teachers, parents, administrators, and students will use them as starting points for an ongoing discussion about classroom activities and curricula.

The primary focus of the standards is on the content of English language arts learning. Content cannot be separated from the purpose, development, and context of language learning. As educators translate these standards into practice, they must consider the unique range of purposes, developmental processes, and contexts that exists in their communities.

SECTION A

Bloom Sheets related to the standard of
Using Print to Build
Global and Personal Understandings

*Standard 1

Students read a wide range of print and non-print texts to build an understanding of texts, of themselves, and of the cultures of the United States and the world; to acquire new information; to respond to the needs and demands of society and the workplace; and for personal fulfillment. Among these texts are fiction and non-fiction, classic, and contemporary works.

*Descriptions of the 12 standards in this book are adapted from *Standards for the English Language Arts,* by the International Reading Association and the National Council of Teachers of English, Copyright 1996 by the International Reading Association and the National Council of Teachers of English. Reprinted with permission.

VARIETY IS THE SPICE OF LIFE

KNOWLEDGE
Distinguish between fiction and nonfiction books.

ANALYSIS
Compare and contrast a fiction and nonfiction book on the same topic or subject. How are they alike and how are they different?

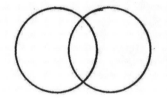

COMPREHENSION
Suggest a good fiction and nonfiction book for your friends to read. Briefly describe each book by writing a brief synopsis of its content.

SYNTHESIS
Design an original book cover for one of the books listed at the Comprehension Level.

APPLICATION
Survey at least ten students in your English class to determine their favorite fiction and nonfiction book to date. Compile the results of the survey.

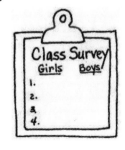

EVALUATION
Consider why some people prefer to read books that are nonfiction rather than fiction for pleasure. Which do you like best and why?

Ignite Student Intellect and Imagination in Language Arts, by Sandra L. Schurr and Kathy L. LaMorte. Published by National Middle School Association, 2007.

CLASSICAL VERSUS CONTEMPORARY WORKS

KNOWLEDGE
How can one tell the difference between a classical text and a contemporary text?

COMPREHENSION
Prepare a list of classical and contemporary books that you would recommend to someone your age. Use the Internet for some insights on this task.

CLASSICAL	CONTEMPORARY
Tom Sawyer	Harry Potter
Little Women	Walk Two Moons
Call of the Wild	Hatchett

APPLICATION
Choose either a classical or contemporary book that you would like to read and READ it!

ANALYSIS
Determine what makes a book a classic.

SYNTHESIS
Pretend you have just written a contemporary novel for teenage readers and it is on the best seller list. Summarize its plot and compose a character sketch of its protagonist.

EVALUATION
Justify reading the classics in middle school.

Ignite Student Intellect and Imagination in Language Arts, by Sandra L. Schurr and Kathy L. LaMorte. Published by National Middle School Association, 2007.

READING A GOOD NOVEL

KNOWLEDGE

Choose a novel to read and record its title, author, publisher, and copyright date. Read your novel and keep a record of the number of pages you cover each day including the starting date and the finishing date of this task.

COMPREHENSION

Draw a map of the setting that includes places where key events took place in the novel or prepare a set of simple illustrations that represent various places in the novel where the action took place.

APPLICATION

Produce a list of five to ten adjectives that best describe the protagonist and the antagonist in the novel or construct a chart that compares and contrasts these two characters.

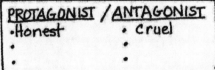

PROTAGONIST / ANTAGONIST
- Honest   • Cruel
- •
- •

ANALYSIS

Analyze the plot of the novel and determine which events can be classified as rising action, conflict, climax, and falling action or make a timeline of the key events in the story putting them in chronological order.

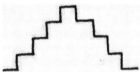

SYNTHESIS

Pretend you are making a movie of the novel and propose what current movie actors/actresses should play the important roles and why. Then, decide on a song that best represents the plot and theme of the novel and explain your rationale.

EVALUATION

Determine your criteria for a good novel and then conclude how this novel measures up against those criteria.

Ignite Student Intellect and Imagination in Language Arts, by Sandra L. Schurr and Kathy L. LaMorte. Published by National Middle School Association, 2007.

SOME KEY GENRES OF LITERARY NOVELS

KNOWLEDGE

Use the glossary of your English textbook to define the concept of "genre" as it relates to literature. Record the definition as written.

COMPREHENSION

Give several examples of the different genres for literary readings and a specific book title for each category.

APPLICATION

During this school year, make it a goal to read at least one novel in each of the identified genres of literature. To get you started, construct a "plan of action" specifying which genres you plan to read in any given period of time. Try to include possible book titles for each genre that you could consider reading as part of your planning process.

ANALYSIS

Which of the genres listed at the Comprehension Level best describe the kind of person you are and the type of genre you prefer? Be specific in your comments.

SYNTHESIS

Create a mini-poster or smart chart that lists, describes, and gives examples of the different genres of literature that you have discovered. Keep this in your notebook as a reference and reminder when you need it.

EVALUATION

If you were to become an author of teenage novels, which type of genre would you want to specialize in and why? Create a BOOK AWARD that might be given to you as an adult writer being recognized for this honor.

Ignite Student Intellect and Imagination in Language Arts, by Sandra L. Schurr and Kathy L. LaMorte. Published by National Middle School Association, 2007.

HISTORICAL EXPRESSIONS FROM THE PAST

KNOWLEDGE

Recite an expression you know from literature. Consider mythologies, fables, biblical stories/verses, or historical events.

COMPREHENSION

Account for each of these literary expressions. Where did they originate? Use the Internet to help you find the source.

(1) Achilles' heel;
(2) Baker's dozen;
(3) Cry wolf;
(4) Drop in the bucket;
(5) Eat humble pie;
(6) Go hog wild; and
(7) Doubting Thomas

APPLICATION

Use each of the literary expressions from the Comprehension Level in an appropriate sentence that demonstrates your understanding of its meaning.

ANALYSIS

Determine how expressions like these differ from idioms. Give two or three examples to illustrate your point.

It's raining cats and dogs!

SYNTHESIS

Create a series of humorous drawings to show the literal, not interpretive, meanings of these expressions.

EVALUATION

Do literary expressions and idioms reflect a specific culture or do they apply to many cultures? Justify your position.

Ignite Student Intellect and Imagination in Language Arts, by Sandra L. Schurr and Kathy L. LaMorte. Published by National Middle School Association, 2007.

SECTION B

Bloom Sheets related to the standard of
Using Varied Genres of Literature

Standard 2

Students read a wide range of literature from many periods in many genres to build an understanding of the many dimensions (e.g. philosophical, ethical, aesthetic) of human experience.

CLASSICAL	CONTEMPORARY
Tom Sawyer	Harry Potter
Little Women	Walk Two Moons
Call of the Wild	Hatchett

A VISUAL PORTRAYAL OF A GOOD NOVEL

KNOWLEDGE

Write down the names and authors of several novels that you have read or plan to read this year. How do the illustrations or graphics on the covers of books grab your attention or give you information about the book's content?

COMPREHENSION

Summarize several ways to report on a book that you have chosen to do or that have been assigned to you by a teacher over the years.

REVIEW OF . . .

APPLICATION

Try one or more of these visual book reporting strategies for a novel that you are reading now or have just completed: (1) Drawing, map, or series of illustrations depicting the setting; (2) Character attribute charts or webs; (3) Scope and sequence ladder or diagram of the plot showing the key events in chronological order; and (4) Collage or paper quilt of novel's theme that includes colors, symbols, images, patterns, and statements depicting the theme.

ANALYSIS

It has been said that "a picture is worth a thousand words." Why are visual representations of people, places, and events sometimes more effective than descriptive passages? Give examples to support your argument.

SYNTHESIS

Create a storyboard that tells the overall story of a novel you have read. Cut apart the storyboard boxes/squares and see if a friend can put them in the correct sequence.

EVALUATION

It has been said that you "can't judge a book by its cover." Do you agree with this statement? Why or why not?

Ignite Student Intellect and Imagination in Language Arts, by Sandra L. Schurr and Kathy L. LaMorte. Published by National Middle School Association, 2007.

FAIRY TALES ON TRIAL

"The better to see you my dear."

KNOWLEDGE

Locate several fairy tales and write down their titles and the culture they represent.

COMPREHENSION

In your own words, explain the fairy tale as a literary genre.

ANALYSIS

Critics of fairy tales claim that their content or story line is often too violent for young children to enjoy. Read several fairy tales from different countries and determine whether you agree or disagree with the critics and why.

APPLICATION

Research to find out the role and responsibilities of each of these participants essential to a jury trial: defense lawyers, prosecution lawyers, bailiffs, judges, witnesses for the prosecution and for the defense; and jury members. Choose a fairy tale on which to simulate or stage a mock jury trial with a group of your peers. Who is on trial and for what? What evidence is there of trespassing, damaging property, robbery, breaking the law, fraud, trickery, criminal behavior, kidnapping, or murder?

SYNTHESIS

Take a popular fairy tale and rewrite it with a different twist.

EVALUATION

What can we learn from the reading and enacting of popular fairy tales? What do they teach us about the outcomes or consequences of our behaviors?

Ignite Student Intellect and Imagination in Language Arts, by Sandra L. Schurr and Kathy L. LaMorte. Published by National Middle School Association, 2007.

THE MAGIC OF FOLKLORE

KNOWLEDGE
What is folklore? How do we recognize it? What does it tell us about human experiences?

Lady bug, lady bug
Fly away home.

COMPREHENSION
Briefly describe each of these forms of writing and give a specific example of each one: Myth, Folktale, Legend, Tall Tale, Fable, Fairy Tale, and Nursery Rhyme.

APPLICATION
Produce an original myth, folktale, legend, tall tale, fable, fairy tale, or nursery rhyme of your own.

The Frog Prince

ANALYSIS
Determine what one can learn about the philosophical, ethical, or aesthetic dimensions of human experience from the various types of folklore.

SYNTHESIS
Illustrate your original work from the Application Level. Put your story in a picture book format.

EVALUATION
Some schools and libraries have actually banned selected folklore pieces from their programs because they contain violence and unsavory characters. How might they justify their position?

Ignite Student Intellect and Imagination in Language Arts, by Sandra L. Schurr and Kathy L. LaMorte. Published by National Middle School Association, 2007.

WAYS TO ENJOY AND SHARE A SHORT STORY

KNOWLEDGE
Select a short story to read from your literature book or a book of short stories from the library. Draw a series of illustrations to show the key events in the story. Include a caption labeling each event depicted.

COMPREHENSION
Summarize the action in your story through a one minute book talk.

APPLICATION
Construct a mask of the main character in the story and use it as a prop to deliver a monologue that tells the story from your personal perspective.

ANALYSIS
Reflect on the story and respond to each of these starter statements:
1. This story made me realize that . . .
2. This story made me wonder about .
3. This story made me feel . . .
4. This story made me see . . .
5. This story made me decide . . .
6. This story would appeal to someone who . . .

SYNTHESIS
Make a tape recording of a portion of the short story read aloud that would make someone else want to read it.

EVALUATION
What are your criteria for a good short story? How does this story measure up against your criteria?

Ignite Student Intellect and Imagination in Language Arts, by Sandra L. Schurr and Kathy L. LaMorte. Published by National Middle School Association, 2007.

LEARNING ABOUT FICTION FORMS

KNOWLEDGE
Use a glossary in a literature textbook to define each of these prose story formats: Allegory, Fantasy, Historical Fiction, Horror, Informational Fiction, Mystery, Realistic Fiction, Romance, Science Fiction, and True Adventure.

COMPREHENSION
Visit the school or local library and find an example of each of the fiction forms at the Knowledge Level. Compile an annotated bibliographic entry for each one.

APPLICATION
Use the Internet to collect information about the authors for each of these book titles from the Comprehension Level. Prepare a short biography on each of these authors.

ANALYSIS
Interview several students in your class to determine which types of fiction they enjoy most. Draw conclusions from these interviews and share your results.

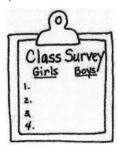

SYNTHESIS
Try composing a story of your own. Create a story map that lists the major characters, that gives descriptive words for potential settings, that includes key words depicting the theme, and that outlines the most important events in the plot. Write a rough draft.

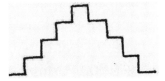

EVALUATION
Rank order the story forms listed at the Knowledge Level in terms of their appeal to you with 1 being your most favorite and 10 being your least favorite. Give reasons for your first and last choices.

Ignite Student Intellect and Imagination in Language Arts, by Sandra L. Schurr and Kathy L. LaMorte. Published by National Middle School Association, 2007.

LEARNING ABOUT NONFICTION FORMS

KNOWLEDGE

Answer this question: How does nonfiction writing differ from fiction writing?

COMPREHENSION

Briefly summarize the purpose for each of these nonfiction forms of writing: Autobiography, Biography, Essay, History, Journal, News Story, Reference, and Travelogue.

APPLICATION

Visit the school or local library and locate examples of each of these types of nonfiction forms of writing. Prepare an annotated bibliographic entry for one example of each form listed at the Comprehension Level.

ANALYSIS

Compare and contrast the procedure for researching the title of a nonfiction book on the Internet with that of using the library for this purpose. How are they alike and how are they different? Which do you prefer and why?

SYNTHESIS

Create a time line of the important events in your life to date. Then, use this information to write an original autobiography of your life.

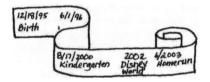

EVALUATION

A book review is a cross between an essay and a report. Write a review of one of the books from the Comprehension Level. Critique its factual information and then embellish it with your opinions of the content and writing style of the author.

Ignite Student Intellect and Imagination in Language Arts, by Sandra L. Schurr and Kathy L. LaMorte. Published by National Middle School Association, 2007.

THE HISTORICAL EVOLUTION OF DRAMA

KNOWLEDGE
Record the dictionary definitions of "drama" and its related terms. Note what they all have in common.

COMPREHENSION
Research the history of drama beginning with the Greeks. Prepare a set of notes on your findings.

APPLICATION
Investigate each of these early forms of drama: Passion Plays, Shakespearean Theater, Chinese Theater, and Japanese Theater.

ANALYSIS
Discover what is unique about each of these forms of drama: Mime, Pantomime, and Opera.

SYNTHESIS
Create an original script for a play. Be sure to assign parts, write down the specific dialogue for each character, and write out stage directions that explain the setting, mood, props, sets, and costumes for the scenes.

EVALUATION
Ludwig Lewisohn (1915) once said: "In all ages the drama, through its portrayal of the acting and suffering spirit of man, has been more closely allied than any other art to his deeper thoughts concerning his nature and his destiny." What is Lewisohn saying about drama in this quote and do you agree with him? Why or why not?

Ignite Student Intellect and Imagination in Language Arts, by Sandra L. Schurr and Kathy L. LaMorte. Published by National Middle School Association, 2007.

IDIOMS MAKE FOR GOOD READING

KNOWLEDGE
Answer this question: What is an idiom?

COMPREHENSION
Give examples of several common idioms that you have heard, used, or read recently.

APPLICATION
Choose any three of these idioms and cite a personal observation, experience, or situation to illustrate the application of the expression:
1. She bawled her eyes out.
2. He's (or she's) on top of the world.
3. She's as neat as a pin.
4. I'm in (or was in) a pretty pickle
5. I'll stay until the bitter end.
6. I just had to blow off steam.
7. He/She has too many irons in the fire.
8. He cried wolf too many times.
9. She's as mad as a wet hen.
10. I was furious, but I held my tongue.

I'm so hungry...
I could eat a horse!

ANALYSIS
Speculate on the origin of one or more of the common idioms you know or have used on a regular basis. What type of historical, social, emotional, or physical event might have triggered its origin?

It's raining cats and dogs!

SYNTHESIS
Create a series of drawings to illustrate the "ridiculousness" of one or more of the idioms from the Application Level.

I'm in a pretty pickle

EVALUATION
Assess how idioms provide humor or comedy in our everyday lives through the reading, writing, speaking, and listening of them.

Ignite Student Intellect and Imagination in Language Arts, by Sandra L. Schurr and Kathy L. LaMorte. Published by National Middle School Association, 2007.

FIVE FANTASTIC POETRY FORMS FOR YOU TO EXPLORE

KNOWLEDGE

Briefly identify and define each of these five poetry forms: Acrostic, Cinquain, Clerihew, Haiku, and Tanka.

Personal
Oblique
Entertaining
Troubador
Refreshing
Yearn

COMPREHENSION

Use a textbook, reference book, poetry book, or the Internet to locate a good example of each of these poetry forms. Copy them down and illustrate them-----one per page, please.

APPLICATION

Construct an original Acrostic, Cinquain, Clerihew, Haiku, or Tanka poem of your own. Illustrate it as well.

Trees in winter time
Heavy with snow branches low
Waiting for spring rains.

ANALYSIS

Use a Venn Diagram to compare the Cinquain, Haiku, and Tanka forms of poetry. Note similarities and differences.

cinquain   Haiku

tanka

SYNTHESIS

Create your own booklet of short, but clever poems that you enjoy. Browse through several poetry books in the library or on line and copy ten of them in your best handwriting. Be sure to include the author and source for each poem. Memorize one or more of them and be ready to have an "informal poetry reading" with several of your peers.

EVALUATION

Select a poem from a documented source that you enjoy and read it thoughtfully. Then answer these questions: (1) What was the poem about? (2) What poetry form does it represent? (3) What were your favorite lines in the poem? (4) How does the poem make you feel?
(5) What message is this poet saying to the reader?

Ignite Student Intellect and Imagination in Language Arts, by Sandra L. Schurr and Kathy L. LaMorte. Published by National Middle School Association, 2007.

MORE GREAT POETRY FORMS FOR YOU TO EXPLORE

KNOWLEDGE
Find an example of each of these special poetry forms: Ballad, Epic, and Sonnet. Read and record the poems on tape.

COMPREHENSION
Briefly differentiate between a ballad, epic, and sonnet poetry form. What do they all have in common?

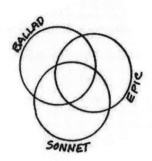

APPLICATION
Research to find out something about each of these early poets of long ago and their contributions to the genre of poetry as we know it today: Shamans, Druids, and Bards.

ANALYSIS
Decide which of the poetry forms listed at the Knowledge Level is basically a sad poem, a romantic poem, and a heroic poem. Which do you prefer and why?

SYNTHESIS
Propose a modern or classical piece of music that you feel best represents the mood of a ballad, the mood of an epic, and the mood of a sonnet. Document your choices in some way.

EVALUATION
What special qualities, writing skills, and personal experiences does it take to be a quality poet?

Ignite Student Intellect and Imagination in Language Arts, by Sandra L. Schurr and Kathy L. LaMorte. Published by National Middle School Association, 2007.

FUN WITH LIMERICK AND CONCRETE POETRY FORMS

KNOWLEDGE

Write down the unique characteristics of both a limerick and a concrete poetry form.

COMPREHENSION

Explain the appeal of both limericks and concrete poems to the reader.

APPLICATION

Words take a special shape in a concrete poem. Make a list of objects that have a distinctive shape such as an umbrella, cat, bell, flag, or tree. Choose one of these objects and make a large sketch or outline of it. Next, write down some specific qualities of the object as well as some things that it does and some things that you associate with it. Write out words, phrases, short sentences, and action statements along the outline of your object using varied type size, color, or print fonts. Go around the outline as many times as you want until your message is completed.

ANALYSIS

Locate several examples of limericks and determine the importance of both rhythm and rhyme in this poetry form. Consider the importance of both stressed and unstressed syllables and rhyming patterns in your analysis. What type of mood is most often portrayed in a limerick form?

SYNTHESIS

Try writing a limerick or a concrete poem. Illustrate it if you like. Share it with classmates.

EVALUATION

Many students much prefer reading and writing stories than they do poetry. Why do you think this is so and what could be done to change their attitudes or perceptions?

Ignite Student Intellect and Imagination in Language Arts, by Sandra L. Schurr and Kathy L. LaMorte. Published by National Middle School Association, 2007.

A GOOD SPEECH IS LITERATURE TOO!

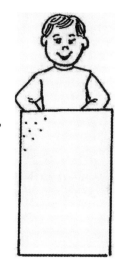

KNOWLEDGE
Locate the words to a given speech from history. Consider such speeches as Abraham Lincoln's "Gettysburg Address," FDR's "The only thing we have to fear is fear itself" speech, or John F. Kennedy's "Ask not what your country can do for you, but what you can do for your country," speech. Read through them carefully.

COMPREHENSION
In your own words, explain what makes these famous speeches so great. Consider their purpose, vocabulary, use of figurative language, historical time period/event, speaker's passion, intended audience, and place where they were given.

APPLICATION
Practice reading one or more of these speeches aloud, perhaps as a choral reading with some of your peers. Assign parts to individuals and small groups. Consider when it is best to pause, when to use voice inflections, when to speak softly/loudly, when to enunciate selected words, and anything else to add drama.

ANALYSIS
It has been said that these famous speeches, along with many others, are speeches of intense passion. Point out the most poignant or passionate passages in one or more of the speeches.

SYNTHESIS
Think of a subject or controversial issue that you feel passionate about such as war, peace, abortion, stem cell research, death penalty, drug testing, same sex marriages, divorce, right to vote, etc. Try writing an emotional speech that conveys your strong feelings on whether something is right or wrong and what should be done about it!

EVALUATION
Research suggests that one of the things people fear most is having to give a speech in front of an audience. Why do you think this is true?

Ignite Student Intellect and Imagination in Language Arts, by Sandra L. Schurr and Kathy L. LaMorte. Published by National Middle School Association, 2007.

SECTION C

Bloom Sheets related to the standard of
Interpreting, Evaluating, and Appreciating Texts

Standard 3

Students apply a wide range of strategies to comprehend, interpret, evaluate, and appreciate texts. They draw on their prior experience, their interactions with other readers and writers, their knowledge of word meaning and of other texts, their identification strategies, and their understanding of textual features (e.g., sound-letter correspondence, sentence structure, context, graphics).

TESTING YOUR TEST-TAKING SKILLS IN ENGLISH

KNOWLEDGE

Define "test anxiety." Tell how it affects you and others when taking an important English test to show what you know and can do in this subject area.

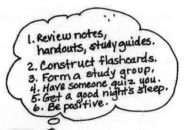

COMPREHENSION

Explain how you prepare for a test. Think about what you do both in and out of class.

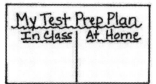

My Test Prep Plan
In Class | At Home

APPLICATION

Compile a list of intelligent ways to use your time wisely when taking a test. Here are two to get you started: (1) Put your name on all pages of the test; (2) Look over the entire test to note the number and types of questions before you begin writing answers to the questions.

ANALYSIS

Determine the advantages and disadvantages of these types of questions for you and others:
(1) True/False; (2)Multiple Choice; (3) Matching; (4) Short Answer; (5) Fill-in-the-Blank; and (6) Essay

SYNTHESIS

Create a humorous comic strip, skit, essay, short story, or monologue on the topic of "HOW NOT TO TAKE A TEST!"

EVALUATION

Rank order these reasons for poor performance on tests as they relate to your personal test-taking skills with 1 being your greatest weakness and 5 being less of a problem for you: (1) Didn't budget enough time to study; (2) Didn't read assigned text pages or do homework; (3) Test anxiety; (4) Didn't check answers and/or left too many answers blank; and (5) Didn't get enough sleep night before or eat breakfast on test day. What will you do differently next time?

Ignite Student Intellect and Imagination in Language Arts, by Sandra L. Schurr and Kathy L. LaMorte. Published by National Middle School Association, 2007.

LEARNING TO STUDY AND STUDYING TO LEARN

KNOWLEDGE

The SQ3R method developed by Francis P. Robinson is a very effective study skill. Write down the words represented by the "S, Q, and R" letters. What does the "3" mean?

SQ3R

COMPREHENSION

In a comprehensive paragraph, elaborate on what it means to Survey, Question, Read, Recite, and Review.

APPLICATION

There are several memory strategies for learning important pieces of information. One popular device is the use of Acrostics. This is when you make up a sentence using the first letter of each word in a list or a sequence. For example, the lines of the musical staff, EGBDF, is remembered by the sentence: "Every Good Boy Does Fine." Make up some acrostics of your own for memorizing key words or concepts in your English class such as the parts of speech or the genres of literature

ANALYSIS

Another good learning tool is that of a Mind Map where you organize mental maps from known information and then fill in any missing data. Examine an important section from your literature or grammar textbook and construct a Mind Map that contains all major terms and concepts for that excerpt.

SYNTHESIS

Invent a new type of graphic organizer that works for you when studying for a poetry test, a grammar test, a story/novel test, or a spelling test.

EVALUATION

Assess your personal study or learning skills by answering these questions and qualifying your answers with specific examples:

(1) Are you organized?
(2) Do you know how you learn best?
(3) Do you know what distracts you?
(4) Do you have a study system for textbooks? And
(5) Do you make good use of your time and mind?

Organization?
Learning style?
Distractions?
Study system?
Use of time? mind?

Ignite Student Intellect and Imagination in Language Arts, by Sandra L. Schurr and Kathy L. LaMorte. Published by National Middle School Association, 2007.

THE CHALLENGE OF WORDS

KNOWLEDGE
Define homophone, homonym, and homograph. Write down an example of each one.

COMPREHENSION
In your own words, explain what is meant by a palindrome and give several examples.

APPLICATION
Acronyms are words formed from the first letters or syllables of words in phrases or titles. Compile a list of ten to twenty different acronyms.

SCUBA

ANALYSIS
Contrast a synonym with an antonym by using them both in the same sentence. Example: Sandy was absent from the church service today but her parents were present for the bible study class.

SYNTHESIS
Compose a series of sentences that contain some of the most commonly misspelled words of students your age. Use the incorrect spelling in each sentence and ask a friend to pick out the word that needs correcting and correct it!

EVALUATION
1. Summarize the five basic rules for spelling as specified below and judge which one(s) are most difficult for you to apply.
2. Words containing ie or ei
3. Words with a silent or final e
4. Words with a final y
5. Words with a consonant preceded by a vowel.
6. Words that follow the one-plus-one Rule.

Ignite Student Intellect and Imagination in Language Arts, by Sandra L. Schurr and Kathy L. LaMorte. Published by National Middle School Association, 2007.

THE MYSTERY OF WORDS

KNOWLEDGE

Write down ten interesting words at your reading level and identify the prefix, suffix, and root word in each one.

Prefix	Root	Suffix
re	admit	ance
trans	late	able

COMPREHENSION

Give several examples of rhyming words that contain five letters or more.

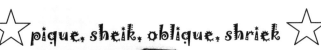

pique, sheik, oblique, shriek

APPLICATION

Compose an effective and grammatically correct sentence for each of these highly descriptive words to demonstrate their meaning: atrocious, brackish, charismatic, dubious, electrifying, futile, gregarious, impetuous, jubilant, kooky, luminous, mystical, nonchalant, offensive, pensive, quirky, remorseful, scrupulous, tenacious, volatile, whimsical.

ANALYSIS

Study these pairs of words to determine why they are words that confuse students. Clarify the definitions for each pair of words:

(1) affect and effect;
(2) complement and compliment;
(3) elicit and illicit;
(4) emigrate and immigrate;
(5) feasible and possible;
(6) principal and principle;
(7) rare and scarce;
(8) reluctant and reticent;
(9) specific and particular;
(10) stationary and stationery.

SYNTHESIS

Compile a set of unusual and descriptive words for each of these categories: poetic words, nature words, feeling words, advertising words, movement words, action words, and sports words.

EVALUATION

Assess several of your writing assignments from the last few days/weeks in terms of your spelling and word selections. On a 1 to 10 scale, rate your overall vocabulary/ spelling skills with 1 being "poor" and 10 being "outstanding." Defend your rating with specific examples from your papers.

Ignite Student Intellect and Imagination in Language Arts, by Sandra L. Schurr and Kathy L. LaMorte. Published by National Middle School Association, 2007.

INITIALS, ACRONYMS, AND ABBREVIATIONS

KNOWLEDGE

Answer this question: When are you likely to encounter initials and abbreviations in your reading as substitutes for the whole word?

Ms. Mary Smith MD PA
389 Maple Ave.
Nokomis, FL 34275
USA

COMPREHENSION

Convert each of these initials into their full expressions or names for things: A.D., ASAP, CD-ROM, C.O.D., CIA, D.A., ERA, FBI, IRS, NAACP, MIA, N.B.A., MTV, POW, R.I.P., RSVP, SAT, UFO, and UN.

APPLICATION

Abbreviations are used to speed up your writing and usually require a period at the end. Abbreviations are never used as words by themselves but as part of a sentence. Write out the word or phrase for each of these abbreviations and use each one properly in a sentence: anon., atty., Ave., Corp., kg., misc., Mt., Pkwy., Rep., Sept., supt., and yds.

ANALYSIS

Analyze each of these acronyms and note how they are all formed and punctuated. Research to find out more factual information about the organization represented by each acronym. They are: MADD, NASA, NATO, NOW, UNESCO, VISTA, WAC, and WHO.

SYNTHESIS

Create a series of fun sentences that include as many initials, acronyms, and abbreviations as you can in each one.

Dr. N.H. Lea, CIA, studies UFOs in AZ.

EVALUATION

Argue against using initials, acronyms, and abbreviations in a written speech, essay, position paper, or formal report.

Ignite Student Intellect and Imagination in Language Arts, by Sandra L. Schurr and Kathy L. LaMorte. Published by National Middle School Association, 2007.

SPELLING WORDS CORRECTLY

KNOWLEDGE
Write down five of the hardest words that you know how to spell and that you think many of your classmates would not know how to spell.

COMPREHENSION
Give examples of at least two words that fit each of these nine rules for spelling English words:

1. When you add "full" to any word, drop the second "l."
2. When you add a "y" or a suffix that begins with a vowel to a word that ends with a silent "e," drop the silent "e."
3. When adding the suffix "ing" to a word that ends with "ie," drop the "e" and change the "i" to "y."
4. When adding a suffix that begins with a consonant (all letters that are not vowels or "y"), do not drop the silent "e."
5. Double the final consonant when you're adding a suffix that begins with a vowel (ing, ed) to a word that ends with a vowel and a consonant (hop, refer).
6. Add "s" to make a word plural if it ends with "y" and the letter in front of the "y" is a vowel.
7. Drop the "y" and add "ies" if the letter in front of the final "y" is a consonant.
8. Add "s" to most nouns that end in "o" to make them plural.
9. "I" before "e" except after "c" or when sounds like "a" as in neighbor and weigh.

APPLICATION
Prefixes are at the beginning of words and when attached to a word, its meaning combines with the meaning of the original word to form a new word. Look up the meaning of each of these Greek and Latin prefixes and then attach them to a second word to form a new word. Prefixes are: arch, contra, dino, hydro, mega, metro, multi, poly, sub, trans.

ANALYSIS
Point out some words you know when the spelling of root words changes in order to receive a suffix.

SYNTHESIS
Create a word finder puzzle that contains some common Greek and Latin roots. Examples: act = activity, action; sphere = hemisphere

EVALUATION
Judge which words are most difficult for you to spell and why.

Ignite Student Intellect and Imagination in Language Arts, by Sandra L. Schurr and Kathy L. LaMorte. Published by National Middle School Association, 2007.

A CRASH COURSE IN SENTENCE STRUCTURE

KNOWLEDGE
Write a declarative sentence, an interrogative sentence, an imperative sentence, and an exclamatory sentence.

COMPREHENSION
Explain what is meant by the simple, complete, and compound subject of a sentence as well as the simple, complete, and compound predicate of a sentence.

APPLICATION
Start with a simple sentence and then make it into a compound sentence and a complex sentence.

1) Jessica fell.
2) Jessica fell and broke her arm.
3) Jessica fell, breaking her arm, and had to be taken to the hospital.

ANALYSIS
Deduce how you can tell when a group of words is not a sentence.

SYNTHESIS
Create a humorous sentence that has both a compound subject and a compound predicate.

Croaking frogs and mating gators create springtime rhapsodies and render the night alive.

EVALUATION
Recommend several ways that one could correct this run-on sentence:

"It might be late before I get home take the cell phone with you so I can call you with details."

Ignite Student Intellect and Imagination in Language Arts, by Sandra L. Schurr and Kathy L. LaMorte. Published by National Middle School Association, 2007.

READING CARTOONS AND COMIC STRIPS

KNOWLEDGE
Locate two editorial cartoons and two comic strips from your local newspaper. Cut them out and use them to complete the following tasks.

COMPREHENSION
Interpret each of the editorial cartoons. What message is each cartoon sending the reader? Classify each cartoon as commentary on a social issue, an economic issue, or a political issue and give reasons for your answer.

APPLICATION
Examine both comic strips. Determine the source of humor for each one. What makes them funny----characters, play on words/deeds, portrayal of human foibles, unusual circumstances, exaggeration, etc.?

ANALYSIS
Determine what comic strips and editorial cartoons have in common?

SYNTHESIS
Create an original comic strip or editorial cartoon of your own.

EVALUATION
Tell whether you agree or disagree with this statement about comic strips and defend why you feel as you do: Comic strips often have very sophisticated and adult-like messages hidden beneath their more obvious and superficial messages. That is why comics appeal to both children and grown-ups alike.

Ignite Student Intellect and Imagination in Language Arts, by Sandra L. Schurr and Kathy L. LaMorte. Published by National Middle School Association, 2007.

THE RELATIONSHIP BETWEEN READING AND WRITING

KNOWLEDGE

How would you answer this question: Are you a better reader or a better writer and how do you know?

ANALYSIS

It has been said that reading and writing are closely linked. How do you think one affects the other? Be specific in your comments.

COMPREHENSION

Summarize the skills you need to be a good writer and the skills you need to be a good reader. Are there any overlapping skills? If so, what are they?

Reader Writer

SYNTHESIS

Choose a favorite author and use the Internet to research biographical information about him/her. Write a letter of appreciation to this author and inquire as to his/her reading habits. Do you get the impression that the author reads as much as he/she writes?

APPLICATION

Keep records of your reading assignments and your writing assignments in school for a week or more. Choose those assignments that you enjoyed the most, that you spent the most time on, and that you did the best job on. Were they more reading-oriented, writing-oriented, or equally dependent upon both your ability to read and write? Explain.

EVALUATION

Determine ways that you could improve both your reading and your writing skills. Can you think of a set of tasks that would improve both simultaneously? Explain.

Ignite Student Intellect and Imagination in Language Arts, by Sandra L. Schurr and Kathy L. LaMorte. Published by National Middle School Association, 2007.

PARTS OF SPEECH: VERBS

KNOWLEDGE
Define verb and be able to tell the difference between a main verb and a helping verb.

COMPREHENSION
Every verb has three main parts called principal parts. Explain when and how each is used.

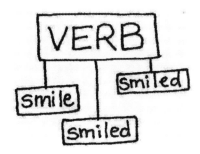

APPLICATION
The tense of a verb is important because it tells you when the action of the verb takes place. There are six main tenses. Write a sentence for each of these six tenses that demonstrates your understanding of how each one is used properly in a sentence.

ANALYSIS
Distinguish between a regular verb and an irregular verb.

SYNTHESIS
Compose a short essay about linking verbs whose title is: "To Be or Not To Be!"

EVALUATION
Defend or negate this statement about verbs: "Whatever you're doing can be expressed by a verb."

Ignite Student Intellect and Imagination in Language Arts, by Sandra L. Schurr and Kathy L. LaMorte. Published by National Middle School Association, 2007.

PARTS OF SPEECH: ADJECTIVES AND ADVERBS

KNOWLEDGE
Define adjective and adverb.

COMPREHENSION
Adjectives answer three questions about the nouns they describe: what kind of? how many? which one? Write three different sentences to show what this means.

One, spotted, stuffed dog remained on the shelf.

Adverbs answer three questions about the verbs which are how, when, and where. Write three different sentences to show what this means.

At dinner time she cooked outside over a fire.

APPLICATION
Demonstrate your understanding of both adjectives and adverbs and where they go in a sentence.

ANALYSIS
Point out how comparisons of adjectives are similar to the way we compare adverbs using positive, comparative, and superlative examples.

SYNTHESIS
Are you more like an adjective

or an adverb? Why?

EVALUATION
Assess ways that adjectives and adverbs make our sentences and paragraphs more expressive, more exciting, and more exacting!

Ignite Student Intellect and Imagination in Language Arts, by Sandra L. Schurr and Kathy L. LaMorte. Published by National Middle School Association, 2007.

PARTS OF SPEECH: NOUNS

KNOWLEDGE
What is a noun? Give several examples.

COMPREHENSION
There are five ways to make regular nouns plural. Describe those five ways and give examples of each. Some nouns are not regular. When they become plural, they either change their spellings or stay the same. Give examples of some irregular plural nouns as well.

APPLICATION
There are eight basic uses of nouns. Construct an original sentence to show each of these applications:
(1) Noun as subject of a sentence;
(2) Predicate (or predicate nominative) noun;
(3) Appositive noun;
(4) Noun as direct object of a verb;
(5) Noun as indirect object of a verb;
(6) Noun as object of a preposition;
(7) Noun as object complement; (8) A possessive noun.

ANALYSIS
Discover the difference between a common noun, proper noun, concrete noun, abstract noun, collective noun, and compound noun.

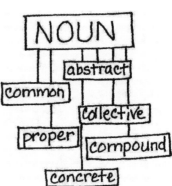

SYNTHESIS
Compose and diagram an original sentence that includes several different types or uses of nouns and is grammatically correct.

EVALUATION
Do you agree or disagree with this statement: "If you know the ways nouns are used, you are less likely to make a mistake using a pronoun."
Give reasons for your answer.

Ignite Student Intellect and Imagination in Language Arts, by Sandra L. Schurr and Kathy L. LaMorte. Published by National Middle School Association, 2007.

PARTS OF SPEECH: PREPOSITIONS, CONJUNCTIONS, AND INTERJECTIONS

KNOWLEDGE
Write down at least five common prepositions, three conjunctions, and three interjections that you like to use in your own writing.

COMPREHENSION
Show how prepositions can tell location, direction, time, and the relationship between a noun or pronoun and another word.

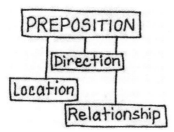

APPLICATION
Construct a series of questions that might appear on a grammar test to show one's understanding of these four kinds of conjunctions:
(1) coordinating conjunctions,
(2) subordinating conjunctions,
(3) correlative conjunctions, and
(4) adverbial conjunctions.

ANALYSIS
How would you describe interjections to someone who had never heard of the term? What kind of punch can they add to a sentence?

SYNTHESIS
Most interjections come at the beginning of a sentence. Complete each of these sentences which starts with an interjection:
1. Whoopee! . . .
2. Oops! . . .
3. Yuck! . . .
4. Good Grief! . . .
5. Wow! . . .

EVALUATION
Conclude why it is so hard for someone who doesn't speak English, to learn the language easily.

Ignite Student Intellect and Imagination in Language Arts, by Sandra L. Schurr and Kathy L. LaMorte. Published by National Middle School Association, 2007.

PARTS OF SPEECH: PRONOUNS

KNOWLEDGE
Answer this question: Why are pronouns such handy little words?

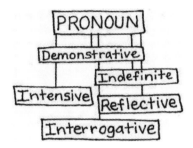

COMPREHENSION
There are seven uses of personal pronouns. Show how pronouns are used to replace nouns:
(1) As subject of a sentence;
(2) As predicate pronoun;
(3) As direct object of a verb;
(4) As indirect object of a verb;
(5) As object of a preposition;
(6) As an appositive; and
(7) To show possession or ownership.

APPLICATION
Find examples of each of these kinds of pronouns in a book that you are reading and write down examples of each:
(1) Demonstrative pronouns;
(2) Indefinite pronouns;
(3) Intensive pronouns;
(4) Reflexive pronouns; and
(5) Interrogative pronouns.

ANALYSIS
Formulate a rule for deciding when to use "who" and "whom" in a sentence. Determine why it is so hard for students to distinguish when to use each of these words correctly.

SYNTHESIS
Devise a handy chart for your notebook that shows the personal pronouns by case, number, and person.

EVALUATION
Assess how one decides whether to use the subject or the object pronoun in a sentence like these:

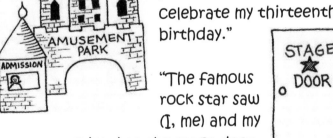

"My parents and (I, me) traveled to an amusement park in another state to celebrate my thirteenth birthday."

"The famous rock star saw (I, me) and my friend at the stage door after attending the concert."

Ignite Student Intellect and Imagination in Language Arts, by Sandra L. Schurr and Kathy L. LaMorte. Published by National Middle School Association, 2007.

SECTION D

Bloom Sheets related to the standard of
Using Spoken, Written, and Visual Language to Communicate

Standard 4

Students adjust their use of spoken, written, and visual language (e.g., conventions, style, vocabulary) to communicate effectively with a variety of audiences and for different purposes.

WRITING ACROSS THE CONTENT AREAS

KNOWLEDGE

Make a list of the times when you must apply your writing skills in math, science, and social studies classes. In other words, what types of writing do you do most in each of these subject areas?

COMPREHENSION

To demonstrate your ability to apply good writing skills in math, try completing one of these activities:
(1) Prepare a one paragraph overview of the metric system; (2) Write an essay describing what the world would be like without numbers; or
(3) Compose a series of word problems for your friends to solve using people and situations you know as the basis for each problem.

APPLICATION

To apply your writing skills in science, try completing one of these activities: (1) In a comprehensive paragraph, differentiate between a crocodile and an alligator or an eagle and a hawk; (2) Prepare a booklet of safety precautions and procedures for dealing with serious weather conditions such as a hurricane; or (3) Develop and perform a skit that teaches others about the solar system.

ANALYSIS

To discover something new on a social studies topic, compare and contrast one pair of these geographical concepts using a graphic organizer:
(1) Weather and Climate; (2) North and South Poles; or (3) Tropical Rainforests and Tropical Grasslands.

SYNTHESIS

Outline the life story of one of these interesting characters:
(1) A sea creature;
(2) A civil rights leader;
(3) A mathematician;
(4) A folklore hero.

EVALUATION

Write an outstanding tribute to one of these individuals whom you admire:
(1) An explorer;
(2) An author;
(3) An astronaut; or
(4) A math teacher.

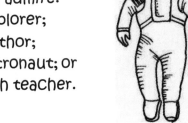

Ignite Student Intellect and Imagination in Language Arts, by Sandra L. Schurr and Kathy L. LaMorte. Published by National Middle School Association, 2007.

THE INTERVIEW PROCESS

KNOWLEDGE
What is the purpose of an interview and what types of interviews are you most familiar with?

COMPREHENSION
Listen to an interview program on television or read an interview type article in a newspaper or magazine. Summarize the overall purpose and content of the interview.

APPLICATION
Set up and conduct an interview with an adult of your choice. Consider a family member,

educator, business person, community leader, local celebrity, or an important person in your life. Establish a time, date, and place for the interview. Consider a telephone interview, a face-to-face interview, or an e-mail interview. Prepare a set of questions to be used ahead of time and plan to tape or record answers as given by the interviewee. Thank the person being interviewed for his/her time and cooperation. Finally, write a summary of your interview results.

ANALYSIS
Do you think it is easier to be the interviewer or the interviewee? Give reasons for your answer.

SYNTHESIS
Stage a mock interview session with a character from the past. Consider interviewing an historical figure, an early scientist, a deceased sports hero, or a past president. Write out a script for the interviewer and the interviewee. Choose a friend to work with you on this task.

EVALUATION
Why are television talk shows that feature interviews as their primary activity so popular with viewers today?

Ignite Student Intellect and Imagination in Language Arts, by Sandra L. Schurr and Kathy L. LaMorte. Published by National Middle School Association, 2007.

WHAT LISTENING TYPE ARE YOU?

KNOWLEDGE

There are three types of listening skills:
(1) Skills showing appreciation;
(2) Skills seeking information; and
(3) Skills wanting to understand and process the information. Recall and share a time or two when you paid attention and listened well for each of these purposes.

COMPREHENSION

In your own words, give examples of things that both kids and adults do which limit their ability to really listen to what others have to say.

APPLICATION

Keep track of your listening skills for a full day or week by coding times when these listening barriers got in your way: Input or information overload; Physical noise or distractions; Personal problems or thoughts; Inattention; or Lack of listening skill training.

ANALYSIS

Debate with a group of peers whether boys listen better than girls, whether kids listen better than adults, whether teachers listen better than students, or whether parents listen better than their children. Draw conclusions from you discussion and analysis process.

SYNTHESIS

Create a mini-poster for your classroom bulletin board or personal notebook of hints or guidelines for good listening habits at school.

EVALUATION

Judge the value of this statement as it relates to the improvement of listening skills: "It is important to listen as much for 'what is not said' as it is for what is said in many circumstances."

Ignite Student Intellect and Imagination in Language Arts, by Sandra L. Schurr and Kathy L. LaMorte. Published by National Middle School Association, 2007.

READER'S THEATER

KNOWLEDGE
During Reader's Theater, students sit or stand and read selected passages from a book, using their oral expression and nonverbal communication skills to create the illusion of drama for the audience. Write down the titles/authors of at least five poems, books, or short stories which you feel would be good for this purpose. Then select one for the set of tasks that follow.

COMPREHENSION
The key to a good Reader's Theater is to read the section of a book, story, or poem that has a large amount of dialogue directly related to the plot/events of the story. Explain why this is effective?

APPLICATION
Reader's Theater should use a narrator to introduce the poem, book, or story and to summarize what has happened in the story line up to the point when the dialogue focuses on the who, what, when, where, and why of the story. Develop a script for the Reader's Theater that you are going to do.

The script should include the words for the narrator's introduction, summary, and interim comments for maintaining the flow of the readings. The script should also include the names of each reader participant and the assigned section or dialogue parts to be read by each individual reader.

ANALYSIS
In Reader's Theater, costumes, props, and scenery are to be avoided. Infer why this is important.

SYNTHESIS
Arrange the presentation of this Reader's Theater for your class. Readers who know their lines well and when not speaking should turn their backs to the audience or look down so as not to distract the audience from the others who are speaking.

EVALUATION
Judge the value of Reader's Theater from an author's perspective, a student's perspective, a reader's perspective, and the perspective of someone in the audience.

Ignite Student Intellect and Imagination in Language Arts, by Sandra L. Schurr and Kathy L. LaMorte. Published by National Middle School Association, 2007.

LET'S GIVE A SPEECH

KNOWLEDGE

List some times or occasions in your life, now or in the future, when it might be very important for you to give a speech of some type.

COMPREHENSION

Prepare to write and give a speech. Choose a strong topic or theme for your speech and plan to make no more than five main points. Use short, powerful, memorable statements to introduce these main points.

APPLICATION

All speeches should follow this organizational structure.
(1) Title and Introduction to grab the audience's attention;
(2) Objective and Overview that outlines the main points you will make.
(3) Body that makes your argument;
(4) Summary that reviews the five main points.
Think of a title and an introduction to get you started. Then, begin gathering some items to support your content such as quotations, facts/figures/statistics, narratives, definitions, humorous jokes, personal anecdotes, current events, or personal experiences.

ANALYSIS

Analyze each of these body structures to use as the foundation for your speech:
(1) Chronological structure (follow a time line;)
(2) Five questions (who, what, when, where, why);
(3) Order of importance (move from least to most important points);
(4) Cause and effect (show result, then explain the process from cause to effect);
(5) Problem-solution (pose a problem and then offer a solution);
(6) Motivational (establish need for your audience and then satisfy that need).

SYNTHESIS

Write, rehearse, and then give your speech in front of a live audience!

EVALUATION

Develop a rubric to assess the quality of your ideas and the delivery of your speech.

Ignite Student Intellect and Imagination in Language Arts, by Sandra L. Schurr and Kathy L. LaMorte. Published by National Middle School Association, 2007.

PUBLIC SPEAKING TIPS AND TIDBITS

KNOWLEDGE
What is an impromptu speech? When might you want to speak to a group in this manner?

COMPREHENSION
Outline some Do's and Don'ts for giving an impromptu speech. Think about vocal and breathing techniques, about eye contact, about word selection, about body language, about pacing your ideas, and about jotting down a few notes to organize your thoughts.

APPLICATION
Prepare 20 – 30 slips of paper with several topics for impromptu speeches including those listed here. Draw a slip out of a hat and practice giving an impromptu speech in front of a group of peers.

Some ideas to get you started are:

(1) It is better to be popular than smart.

(2) Explain what you think makes a test/exam fair.

(3) Say something about freedom.

(4) Explain how you feel about war.

(5) Explain what is good/bad about this country.

ANALYSIS
Which type of impromptu speech do you think would be most difficult for you to give-----an informative speech to teach or  provide new information on a topic; a persuasive speech to convince or influence your audience to agree and/or to act; an entertaining speech to amuse or bring people together.

SYNTHESIS
Think of ways you might deal with nervousness when getting ready to give a speech of any type. What gives you confidence?

EVALUATION
Pretend you are the classroom teacher who has just heard a group of students give their impromptu speeches. List the criteria you think he/she should use in grading the quality and delivery of these speeches.

Ignite Student Intellect and Imagination in Language Arts, by Sandra L. Schurr and Kathy L. LaMorte. Published by National Middle School Association, 2007.

GIVING A MINI-NEWSPAPER SPEECH

KNOWLEDGE

Select a newspaper article of special interest to you from the local paper or downloaded from the Internet.

COMPREHENSION

Prepare a set of five note cards to use with your speech. Record name and date of newspaper as well as the name of news service and reporter of article on card one. Review article by telling who, what, when, where, and how of events in article on card two. On card three, explain why the article is important and what you and others can learn from it. On card four give your reactions to the article, and on card five speculate on how the results of this event might ultimately affect your life.

APPLICATION

Give your newspaper mini-speech to a small or large group of friends using your note cards as prompts when needed.

ANALYSIS

After your speech, write a paragraph on one of these topics:
(1) What might a follow-up story be about?
(2) Who is most affected by this event/issue?
(3) Would you have liked to be a part of this event or issue?
(4) What other current events/issues are related to this one?

SYNTHESIS

Publish a classroom newspaper complete with news/feature stories, editorials, book/movie reviews, comic strips, etc.

EVALUATION

Assess the effectiveness of your mini-newspaper speech using these criteria: Choice of article, Quality of information, Completeness of note cards, and Delivery of speech.

Ignite Student Intellect and Imagination in Language Arts, by Sandra L. Schurr and Kathy L. LaMorte. Published by National Middle School Association, 2007.

FUN WITH CHORAL READINGS

KNOWLEDGE

A choral reading is a type of unison reciting whereby students read a selection aloud together striving for fluidity in their voices. The key to successful choral reading is to blend the group's voices so they sound as if they are one. Why would it be fun to do a choral reading in your classroom?

COMPREHENSION

A choral reading can be centered around a poem, lyrics of a song, essay, short story, famous speech, historical document, or a hymn. Collect several possible springboards that you might consider using as a basis for a choral reading and explain what appeals to you about each of them.

WHERE the SIDEWALK ENDS

Shel Silverstein

APPLICATION

Make a final selection of a piece for the choral reading and then divide it into several parts or sections. You can have solo parts 1, 2, or 3. You can have small group parts A, B, and C. You can have Boy and Girl parts. You can have all parts or combinations of these options.

ANALYSIS

Choral readings can be enhanced through use of props, costumes, sound effects, background music, physical gestures, and movement. Analyze your selection for the choral reading and think of ways that you might enhance its presentation to an audience.

SYNTHESIS

Rehearse your choral reading several times and present it to an audience.

EVALUATION

Assess the effectiveness of the choral reading by completing these starter statements:

(1) My ability to do choral reading is . . .

(2) If I heard someone say "choral reading is fun," I'd respond . . .

(3) Choral reading is a creative activity because. . .

(4) I would describe the quality of our performance as . . . because . . .

Ignite Student Intellect and Imagination in Language Arts, by Sandra L. Schurr and Kathy L. LaMorte. Published by National Middle School Association, 2007.

SECTION E

Bloom Sheets related to the standard of
Using a Range of Strategies for Writing

Standard 5

Students employ a wide range of strategies as they write and use different writing process elements appropriately to communicate with different audiences for a variety of purposes.

MULTIPLE PURPOSES FOR WRITING

KNOWLEDGE

It is important to know the nature of the audience one is writing for. How might the audience affect the subject matter? How might the audience affect the words you use and the writing style you choose? How might the audience affect the way you organize your information?

COMPREHENSION

Distinguish between these four types or purposes for writing: Expository or Explanatory, Descriptive, Narrative, and Persuasive.

APPLICATION

Examine each of these organizational frameworks as a possible way to organize one's ideas in a piece of writing. Discuss each framework in a short paragraph. Frameworks to consider are: Chronological, Order of Importance, Cause-Effect, Problem-Solution, Five Questions (Who, What, When, Where, How), Comparison and Contrast, Opinion and Supporting Evidence, Main Idea and Details.

ANALYSIS

Formulate an opinion about which framework from the Application Level would work best for you when writing an essay, a research paper, a book report, an editorial, a letter of complaint or advice, a short story, and a position paper.

SYNTHESIS

Choose an audience, purpose for writing, and organizational framework from those suggested at the previous levels and compose an original piece of writing.

EVALUATION

Develop a rubric of criteria for assessing your original piece of writing. Use it to proof and critique your final masterpiece!

Ignite Student Intellect and Imagination in Language Arts, by Sandra L. Schurr and Kathy L. LaMorte. Published by National Middle School Association, 2007.

MORE ON THE PURPOSES FOR WRITING

KNOWLEDGE

There are four common types of paragraphs to consider in your writing and they depend on your overall purpose for writing. Define each of these paragraph options: Descriptive, Narrative, Expository, and Persuasive.

COMPREHENSION

Review many paragraphs in essays, short stories, newspaper/magazine articles, reports, reviews, or novels. Find and read examples of all four types of paragraphs defined at the Knowledge Level. Then, in your own words, summarize the common and effective elements you found in these paragraphs.

APPLICATION

Think of several good topics for each of the four types of paragraphs. Next, choose a topic and construct a quality paragraph of your choice.

ANALYSIS

Compare and contrast all four of these types of paragraphs. How are they alike and how are they different?

SYNTHESIS

A good paragraph has a topic sentence and a series of related details that support the overall topic or purpose of the paragraph. It is important to arrange your details in the most effective order. Compose an original paragraph on any topic and choose one of the following methods for arranging the supporting details: Chronological order, Order of location, Order of importance, Cause and effect, Comparison and contrast, or Illustration.

EVALUATION

Judge which types of paragraphs are most often found in fiction writing and in non-fiction writing. Which paragraph types seem to be favored in each type of writing?

Ignite Student Intellect and Imagination in Language Arts, by Sandra L. Schurr and Kathy L. LaMorte. Published by National Middle School Association, 2007.

WRITING A FIVE PARAGRAPH ESSAY

KNOWLEDGE

Use these questions to help you select a topic for your essay: (1) What am I interested in writing about? (2) Do I have a special knowledge in a particular area? (3) What do I want to learn about? (4) How much time do I have?

COMPREHENSION

Choose one of these purposes for your essay and explain why it works best for your topic: (1) A narrative essay that tells a story by relating a sequence of events; (2) A descriptive essay that focuses on an event, person, object, or setting and depends upon details and images; (3) An explanatory essay that explains, analyzes, or interprets an issue; (4) An argumentative essay that attempts to persuade readers to take some action or convince them of the writer's position.

APPLICATION

Follow this structure when writing your essay: (1) Introduction: One paragraph that introduces your topic to the reader;

(2) Body: Three paragraphs that present the evidence in an orderly manner; 3) Conclusion: One paragraph that brings a tone of finality to the essay.

ANALYSIS

Select one or more of these strategies as a plan for making your paragraphs come alive:
(1) Set of examples and illustrations to make a point;
(2) Definitions;
(3) Analogies;
(4) Comparisons and contrasts;
(5) Cause and effect situations;
(6) Steps of a process or experience.

SYNTHESIS

Grab the attention of the reader by employing one of these ways to start your essay: (1) Startling statement or statistic; (2) Poignant quotation; (3) Simple case study; (4) Funny anecdote; (5) Provocative question.

EVALUATION

Construct a rubric for evaluating your essay? What criteria will you use? What rating scale works best? How will you solicit feedback from readers?

Ignite Student Intellect and Imagination in Language Arts, by Sandra L. Schurr and Kathy L. LaMorte. Published by National Middle School Association, 2007.

CREATIVE WRITING TECHNIQUES FOR YOU TO CONSIDER

KNOWLEDGE
Writing techniques are special ways of treating words when you are engaged in creative writing. Define each of these specific techniques: Caricature, Exaggeration, Flashback, Foreshadowing, Irony, and Monologue.

Pecos Bill

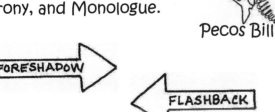

COMPREHENSION
Using your favorite short stories or novels, locate and record an example of each of the six techniques listed at the Knowledge Level.

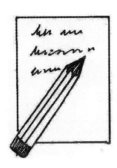

APPLICATION
Try writing an original version of one or more of these six techniques during a writing assignment.

ANALYSIS
Determine how each of the six techniques can enhance the quality of a story for the reader. Be able to share specific examples to document your conclusions.

SYNTHESIS
Create a simple picture or diagram that illustrates each of these six techniques through drawings or sketches rather than words.

EVALUATION
Who is a favorite author of yours because he/she uses literary techniques to enhance the elements of the story line? Defend your position.

Ignite Student Intellect and Imagination in Language Arts, by Sandra L. Schurr and Kathy L. LaMorte. Published by National Middle School Association, 2007.

MORE WRITING TECHNIQUES FOR YOU TO CONSIDER

KNOWLEDGE
Define each of these creative strategies for enhancing words on a page and engaging the reader: Similes, Metaphors, Personification, Assonance, Imagery, Hyperbole, Allusions, Alliterations, and Onomatopoeia.

COMPREHENSION
Locate specific examples of the writing techniques from the Knowledge Level and write them down for future reference.

APPLICATION
Try writing an original version of one or more of these nine techniques during a writing assignment.

ANALYSIS
Debate whether these writing techniques are more likely to be found in poetry or prose. Provide data for your conclusion.

SYNTHESIS
Design a mini-poster that teaches others about these nine techniques for "jazzing up" one's writing or for "tickling and teasing" one's imagination.

EVALUATION
Which of these creative writing strategies are easiest for you to use and recognize? Most difficult for you to use and recognize? Why do you think that is?

Ignite Student Intellect and Imagination in Language Arts, by Sandra L. Schurr and Kathy L. LaMorte. Published by National Middle School Association, 2007.

A WRITER'S TOOLBOX

KNOWLEDGE
Answer this question: When would a writer want to use a dictionary and a thesaurus?

COMPREHENSION
Part of writing well is knowing how to use the mechanics of writing. Prepare a list of the following:
(1) Rules for using commas; [?]
(2) Rules for using semi-colons; [;]
(3) Rules for using quotation marks; (4) Rules for using dashes; [-]
(5) Rules for using ellipses. Determine how these tools can make you a more interesting and knowledgeable writer. Keep these lists in your notebook as a learning tool.

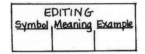

EDITING		
Symbol	Meaning	Example

APPLICATION
Collect information about basic editing symbols and construct a chart that has three columns. Column one should have the editing symbols; column two should have the meanings of these symbols; and column three should have an example of each of these symbols being used to correct an error in a sentence. Keep this chart in your notebook as a learning tool.

ANALYSIS
Debate the difficulty of writing prose versus poetry; of writing a biography versus an autobiography; of writing a comedy versus a tragedy.

SYNTHESIS
Make a list of all the words, phrases, and images that you think of when you hear the word "freedom." Then write a draft paragraph or two using some of those words/phrases/images describing your thoughts, emotions, and ideas about what freedom means to you.

EVALUATION
Proof your paragraphs and use the editing symbols as part of the proofing process. Correct your mistakes and use a dictionary or thesaurus to enhance some of the imagery and descriptive words in your writing.

Ignite Student Intellect and Imagination in Language Arts, by Sandra L. Schurr and Kathy L. LaMorte. Published by National Middle School Association, 2007.

GRAPHIC ORGANIZERS

KNOWLEDGE
What is a graphic organizer and what is its purpose?

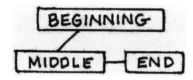

COMPREHENSION
Some common graphic organizers are the web, Venn diagram, wheel, ladder, flow chart, story map, fishbone, cause/effect chart, and KWL chart.

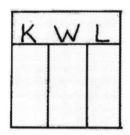

APPLICATION
Select a graphic organizer and use it to record information about something you are reading about or studying in one of your classes. Try using several different organizers for this purpose so that your study notes in actuality become your graphic notes!

ANALYSIS
Determine which type of learner-----visual, auditory, or kinesthetic (tactile) learner-----would be most comfortable with the graphic organizer as a study aid and why.

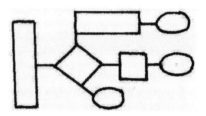

SYNTHESIS
Invent a new graphic organizing structure that makes sense to you and use it!

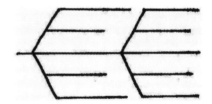

EVALUATION
Students who think on a more abstract (rather than concrete) level like to put things into words, not pictures. Which works best for you and how do you know?

Ignite Student Intellect and Imagination in Language Arts, by Sandra L. Schurr and Kathy L. LaMorte. Published by National Middle School Association, 2007.

DIG THAT DICTIONARY!

KNOWLEDGE
Record the title, publisher, and copyright date of the dictionary that you use most in your school setting.

COMPREHENSION
Find several examples of interesting words in the dictionary that you might use in a creative writing assignment. Write these main entry words down along with their guide

words, correct pronunciations, parts of speech, word divisions, and multiple definitions.

APPLICATION
Use each of the words from the Comprehension Level in a complete sentence.

ANALYSIS
Compare and contrast a dictionary with a thesaurus. How are they alike and how are they different?

SYNTHESIS
Assign a dollar amount to each letter of the alphabet with A = $1.00, B = $2.00, C = $3.00 etc. Find the value of each word listed at the Comprehension Level. Find the value of your complete name.

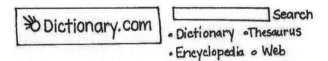

EVALUATION
Visit a dictionary/ thesaurus site on the Internet. Prepare a commentary on how these electronic reference sources compare with those in print. Which do you prefer and why?

Ignite Student Intellect and Imagination in Language Arts, by Sandra L. Schurr and Kathy L. LaMorte. Published by National Middle School Association, 2007.

PERSONAL WRITING

KNOWLEDGE
Look up the dictionary definitions for diary, journal, and log. Write them down. Think about a situation when you would use each type of writing.

COMPREHENSION
Keep a daily diary of your personal activities, thoughts, and feelings for one week. Make each entry as detailed as time and space allow.

APPLICATION

Choose a famous quotation, an editorial cartoon, a comic strip, a set of lyrics for a song, a magazine illustration, a short poem, and a newspaper article. Use each of these items as the focus for a personal journal response entry. Paste each item on a separate piece of paper and write your journal response below.

ANALYSIS
Determine what types of work, people, events, or situations would best be represented by maintaining a log.

SYNTHESIS
Imagine that you are famous and someone has paid a lot of money to purchase your memoirs through the contents of a personal diary, journal, or log kept over time. Compose a sample entry about something special you did that was extraordinary!

lewis and Clark's Journal of Discovery 1803-1806

EVALUATION
What are some long and short term benefits that a person your age might receive from maintaining a personal diary, journal, or log over time.

Ignite Student Intellect and Imagination in Language Arts, by Sandra L. Schurr and Kathy L. LaMorte. Published by National Middle School Association, 2007.

WRITING ADVERTISING COPY

KNOWLEDGE

Browse through several magazines and newspapers looking for different sizes and types of advertising copy. Try to collect 10 to 15 different ads.

COMPREHENSION

Classify or group your collection of ads in some way. Focus on the content and choice of words in the message as well as on the overall graphics and advertising strategy used.

APPLICATION

Imagine you work in the advertising department of a publishing house for a series of teenage novels. Complete an advertising blurb that you could put on the back cover promoting the series.

ANALYSIS

Analyze the collection of ads from the Knowledge Level and redesign one of them that you think needs improving.

SYNTHESIS

Prepare a classified ad that is looking for someone to write dynamic and creative advertising blurbs for a book publisher. What special skills and unique experiences are needed? What are the working conditions and the benefits?

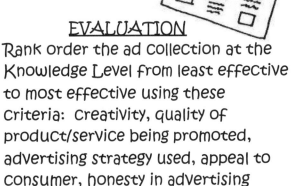

EVALUATION

Rank order the ad collection at the Knowledge Level from least effective to most effective using these criteria: creativity, quality of product/service being promoted, advertising strategy used, appeal to consumer, honesty in advertising message. Give reasons for your first and last choices.

Ignite Student Intellect and Imagination in Language Arts, by Sandra L. Schurr and Kathy L. LaMorte. Published by National Middle School Association, 2007.

AN OVERVIEW OF POETRY TECHNIQUES

KNOWLEDGE
Poetry is a very concentrated form of writing which is imaginative, emotional, thought-provoking, and introspective. Locate a poem that you enjoy and that you feel typifies each of these characteristics. Copy it down to share with others.

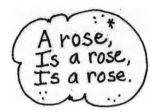

COMPREHENSION
Briefly describe each of these characteristics and patterns of traditional poetry: alliteration, assonance, consonance, end rhyme, foot, meter, internal rhyme, onomatopoeia, quatrain, repetition, stanza, and verse.

APPLICATION

Search for interesting examples of each of these characteristics and patterns of poetry listed at the Comprehension Level.

ANALYSIS
It has been said that of all the forms of creative writing, none is more loved or more hated than poetry." Why do you think this is so?

SYNTHESIS
Pick one or more of these characteristics/patterns of traditional poetry and create an original of your own. Can you write an alliterative sentence or a sentence with repetition of consonant sounds? Can you write lines of poetry with internal rhyme or end rhyme? Can you write a quatrain of four lines or a couplet of two lines?

EVALUATION
What does it take to read and appreciate a poem?

Ignite Student Intellect and Imagination in Language Arts, by Sandra L. Schurr and Kathy L. LaMorte. Published by National Middle School Association, 2007.

POETRY TYPES, TERMS, AND TIDBITS

KNOWLEDGE
Write down the following forms of poetry in alphabetical order: elegy, sonnet, blank verse, limerick, ballad, ode, haiku, cinquain, clerihew, acrostic, tanka, lyric, epic, concrete, and diamante.

Trees in winter time
Heavy with snow branches low
Waiting for spring rains.

COMPREHENSION
Prepare a set of note (flash) cards for all of the poetry types from the Knowledge Level. Add others if you wish. On one side of the card, write the term, its line/syllable count and its definition. On the other side of the card, find an example of the poetry type and copy it down.

APPLICATION
Prepare a series of rhyming word webs. To construct this graphic organizer, draw several big circles on a piece of notebook paper. Write the word you would like to rhyme in the center of the circle.

Place a number of smaller circles around each of the larger ones. Use a dictionary or thesaurus to help you find appropriate rhyming words to fill in the small circles.

ANALYSIS
Locate a narrative, ballad, elegy, epic, lyric, ode, or sonnet poem that you like. Analyze the poem by answering these questions:
(1) What is this poem about?
(2) How does this poem make you feel? (3) What words and lines in the poem have special appeal for you?
(4) What message is the poet trying to get across to the reader?

SYNTHESIS
Try writing an original acrostic, cinquain, clerihew, concrete, diamante, limerick, haiku, or tanka poem of your own.

 ## EVALUATION
Of all the poets you have encountered in your classes over the years, which one is your favorite and why?

Ignite Student Intellect and Imagination in Language Arts, by Sandra L. Schurr and Kathy L. LaMorte. Published by National Middle School Association, 2007.

FUN WITH THINK-A-GRAMS

KNOWLEDGE
THINK-A-GRAMS are a form of lateral thinking. What is lateral thinking?

COMPREHENSION
Here is an example of a THINK-A-GRAM: Figure out what this puzzle is saying: man

board

Hint: It is an expression used when somebody falls out of a boat!

APPLICATION
Figure out these additional THINK-A-GRAMS:

Stand   O   dice
I   M.D. Ph.D   dice

ecnalg   THINK

Answers: I understand; two degrees below zero; pair of dice; backward glance, and Think Big!

ANALYSIS
Why are THINK-A-GRAMS a form of lateral thinking? What makes them so tricky?

SYNTHESIS
Make up several THINK-A-GRAMS of your own and try out on your friends.

EVALUATION
How does one's knowledge of language structure and language conventions help you think of and solve these Think-A-Grams?

Ignite Student Intellect and Imagination in Language Arts, by Sandra L. Schurr and Kathy L. LaMorte. Published by National Middle School Association, 2007.

SECTION F

Bloom Sheets related to the standard of
Applying Knowledge of Language to Create and Critique Texts

Standard 6

Students apply knowledge of language structure, language conventions (e.g., spelling and punctuation), media techniques, figurative language, and genre to create, critique, and discuss print and non-print texts.

READING TO LEARN

KNOWLEDGE
What makes a good reader? Write down the characteristics of someone you know who likes to read and is a good reader.

COMPREHENSION
Explain the KWL technique for reading to learn and a time when you have used it or plan to use it.

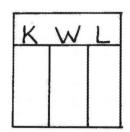

APPLICATION
Apply these five steps in the study-reading process referred to as SQ3R: Survey, Question, Read, Recite, and Review.

ANALYSIS
Another popular technique for learning how to read more effectively is one based on "word pictures." These include everything from mapping to graphic organizers. Search your textbook or the Internet for examples of many different visual aids that fall into these categories. Choose one or more of them to use on a current assignment in class.

SYNTHESIS
There are several types of context clues to help you make sense of new words in a sentence. These are: (1) Clues supplied through definition or association; (2) Clues supplied through synonyms; (3) Clues which appear in a series; (4) Clues derived from cause and effect; (5) Clues contained in comparisons and contrasts; (6) Clues provided by the tone and setting; and (7) Clues through association with other words in a sentence. Compose an original sentence of your own to demonstrate your understanding of each of these context clue applications.

EVALUATION
Tell why you agree or disagree with this statement and give reasons to defend your position: "Readers (and writers) are born and not made."

Ignite Student Intellect and Imagination in Language Arts, by Sandra L. Schurr and Kathy L. LaMorte. Published by National Middle School Association, 2007.

PARTS OF SPEECH THAT SPEAK TO YOU

KNOWLEDGE
Write down the textbook or dictionary definition for each of the eight parts of speech.

COMPREHENSION
Give one good example of each of the following parts and forms of speech through their use in a complete sentence: (1) Forms of nouns: common, proper, concrete, abstract, collective, and compound; (2) Forms of pronouns: personal, possessive, demonstrative, reflexive, intensive, and interrogative; (3) Forms of verbs: action, linking, helping, regular, and irregular; (4) Forms of adjectives: common, proper, and demonstrative; (5) Forms of adverbs: adverbs used with verbs, with adjectives, and with adverbs; (6) Forms of prepositions: one, two, and three-word prepositions; (7) Forms of conjunctions: coordination, subordinating, correlative, and adverbial; and (8) Forms of interjection.

APPLICATION
Practice diagramming several of the above sentences to help you understand the various parts and forms of speech.

ANALYSIS
Study several different sentences on a page of any book you are reading to determine the writing style of the author. Does he/she use lots of descriptive adjectives and adverbs? Does he/she favor certain prepositions over others? Does he/she like to use interjections for emphasis? Does he/she use conjunctions consistently through heavy use of compound and complex sentences? What else do you notice?

SYNTHESIS
Create a GRAMMAR AWARD to give to those people in the class who have mastered the parts of speech in their writing.

EVALUATION
Are you more like a noun or a verb? An adjective or an adverb? A preposition, conjunction, or interjection? How do you know?

Ignite Student Intellect and Imagination in Language Arts, by Sandra L. Schurr and Kathy L. LaMorte. Published by National Middle School Association, 2007.

PUNCHING OUT A PUNCTUATION GUIDE

KNOWLEDGE

On a 3 x 5 file card, write down the correct punctuation mark for each of the following terms: Apostrophe, Colon, Comma, Exclamation Point, Hyphen, Parentheses, Period, Question Mark, Quotation Marks, Dashes, Hyphens, Semicolon, and Underline. Put each mark with its name on one side of the card and save the other side for the following Comprehension Level task.

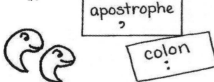

COMPREHENSION

On the other side of each file card, outline the rules for using each punctuation mark correctly. These can now serve as a set of flash cards for learning the information.

APPLICATION

Locate multiple applications of these punctuation rules in newspapers, magazines, and other miscellaneous reading materials.

Cut these examples out, label them, and paste them on additional file cards or in a mini-scrapbook as additional reference tools or learning aids.

ANALYSIS

How does a dash sometimes work like a comma and sometimes like a colon?

SYNTHESIS

Select an interesting essay or short story to read aloud while a group of students hold up the appropriate note (flash) card when the reader comes to that spot in the storyline.

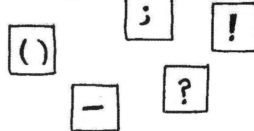

EVALUATION

Evaluate the role punctuation marks play in the structure and interpretation of a sentence. Try to construct a sentence, for example, that has a different meaning if a given mark, such as the comma, appears at a different place in the sentence.

Ignite Student Intellect and Imagination in Language Arts, by Sandra L. Schurr and Kathy L. LaMorte. Published by National Middle School Association, 2007.

SENTENCES THAT RANGE FROM SIMPLE TO COMPLEX

KNOWLEDGE
Use your textbook to record examples of several interesting sentences----one that includes a simple subject and a simple predicate, one that includes a complete subject and a complete predicate, and one that includes a compound subject and predicate.

COMPREHENSION
In your own words, explain the difference between a simple sentence, a compound sentence, and a complex sentence.

APPLICATION
Construct an original declarative sentence, interrogative sentence, imperative sentence, and exclamatory sentence.

ANALYSIS
Do you agree or disagree with these two statements about sentences: "Quality sentences can be short or long, simple or complicated. But all sentences fall into one of three categories----simple, compound, and complex." Defend your position.

SYNTHESIS
Create an original comic strip or short story with one of these titles: The Adverb Who Wanted To Be An Adjective; The Verb That Let The Sentence Down; The Simple Sentence That Became Complex; The Subject Who Didn't Like Its Predicate.

EVALUATION
How does the computer help or hinder students in learning how to write good sentences in their school work?

Ignite Student Intellect and Imagination in Language Arts, by Sandra L. Schurr and Kathy L. LaMorte. Published by National Middle School Association, 2007.

READING AND WRITING BOOK REVIEWS

KNOWLEDGE
Use the Internet to locate Web sites that contain book reviews by both professional reviewers and readers themselves. Download several of these reviews to use with the activities which follow. If possible, find reviews that cover both fiction and nonfiction books.

COMPREHENSION
Discuss the multiple purposes for reading and writing book reviews. What types of information does one expect to find in these reviews? Who reviews these books and how are the books chosen? Who publishes these reviews and who reads these reviews?

APPLICATION
Select a book to read and review. After reading, write out your review for others to enjoy.

ANALYSIS
Infer the many reasons that readers submit online reviews to Web sites such as Amazon.com. How do they differ from the professional reviews online?

SYNTHESIS
Submit a reader review online for the book at the Application Level. How might it differ from your first review? Rewrite it if necessary.

EVALUATION
How much confidence do you have in a review that you read in a newspaper, magazine, or online? Do you place more emphasis on a book review by a professional writer hired to do this kind of work or on a lay reader who does it for fun or because he/she loves to read? Explain.

Ignite Student Intellect and Imagination in Language Arts, by Sandra L. Schurr and Kathy L. LaMorte. Published by National Middle School Association, 2007.

HOW TO BE A BETTER WRITER

KNOWLEDGE
Think about all the different types of writing that you are asked to do in school every day, week, month, or year. Make a personal list of the various writing tasks required of you in class and across content areas.

COMPREHENSION
Explain how you might accomplish each one of these challenges for improving your grammar skill writing in school: (1) How could you "vitalize a verb"? (2) How could you "paint vivid word pictures"? (3) How could you "extend a sentence"? (4) How could you "capture your reader's attention"? (5) How could you "build bridges between words and thoughts"?

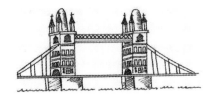

APPLICATION
Determine what you would do to improve your creative writing skills in school if you were told to:
(1) "Spice up your dialogue with colloquialisms or idioms;"

(2) "Jazz up your images with similes and metaphors;"
(3) "Get noisy with onomatopoeia;"
(4) "Allure the reader with alliteration or analogies;"
(5) Bring things to life through personification or personal anecdotes."

ANALYSIS
Formulate an opinion about what literary tools and techniques will help you "tease the minds and tickle the imaginations" of your readers the most!

SYNTHESIS
Write a short essay, story, or poem where you "surprise the reader" with a special literary device when it is least expected!

EVALUATION
Tell why you agree or disagree with this statement: "Writing is a craft."

Ignite Student Intellect and Imagination in Language Arts, by Sandra L. Schurr and Kathy L. LaMorte. Published by National Middle School Association, 2007.

LEADING A BOOK DISCUSSION

KNOWLEDGE
Indicate what this statement means to you: "A book is only half read until you talk about it."

COMPREHENSION
Suggest a set of guidelines for leading an effective book discussion in your classroom.

APPLICATION
Select a book that you have read and prepare a set of quality questions that you would ask others in a student-led book discussion.

ANALYSIS
Would you rather lead a book discussion or limit your role as participant in a book discussion? Give reasons for your answer.

SYNTHESIS
Use the guidelines from the Comprehension Level and the questions from the Application Level and lead a book discussion for your peer group.

EVALUATION
Assess what went well in your book discussion and what you would do differently next time.

Ignite Student Intellect and Imagination in Language Arts, by Sandra L. Schurr and Kathy L. LaMorte. Published by National Middle School Association, 2007.

SECTION G

Bloom Sheets related to the standard of
Generating Ideas and Questions and Posing Problems

Standard 7

Students conduct research on issues and interests by generating ideas and questions and by posing problems. They gather, evaluate, and synthesize data from a variety of sources (e.g., print and non-print texts, artifacts, people) to communicate their discoveries in ways that suit their purpose and audience.

MEDIA CENTER MARVELS

KNOWLEDGE

List all of the print and nonprint resources in the media center that are available to you for conducting research on any school assignment.

ANALYSIS

What are the pros and cons of using print sources of information in the media center rather than online sources of information in completing course assignments.

COMPREHENSION

As school media centers have expanded to meet the needs of students growing up in a technology-dependent society, they have become more complex and more expensive in many ways. Explain how this has influenced the overall school program for helping students to become more efficient and effective researchers.

SYNTHESIS

Build an argument to support the idea that the printed book as a source of information in the media center will be completely replaced by technology within the next few years.

APPLICATION

Determine what tools you are most likely to use in the media center when gathering, evaluating, and synthesizing data from a variety of print and nonprint resources.

EVALUATION

In your opinion, are all of the resources in your school media center used to the full advantage of both students and teachers alike? Defend your position.

Ignite Student Intellect and Imagination in Language Arts, by Sandra L. Schurr and Kathy L. LaMorte. Published by National Middle School Association, 2007.

DATABASES

KNOWLEDGE
A database is a collection of data, facts, statistics, or information. Write down as many different types of databases as you can think of.

ANALYSIS
It has been said that many databases are a threat to our privacy. Formulate arguments to support or negate this idea.

COMPREHENSION
Describe the database programs on the computer that you find most helpful to use when completing homework and in-class assignments. Describe any databases that you have developed for your own personal and academic use.

SYNTHESIS
Pretend you are in charge of creating a database for the next group of students entering your grade next year. Write down possible things you would want to keep track of for this upcoming group of middle school kids.

APPLICATION
Databases take a great deal of time and energy to develop and make operational. Complete a comprehensive paragraph defending this process titled: "The Means Justify the Ends."

EVALUATION
Recommend ways that one can judge the validity of information found on databases that are being used for research or investigative purposes.

Ignite Student Intellect and Imagination in Language Arts, by Sandra L. Schurr and Kathy L. LaMorte. Published by National Middle School Association, 2007.

READING CHARTS, TABLES, AND GRAPHS

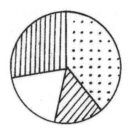

KNOWLEDGE
Find several examples of charts, tables, and graphs in your textbooks or in newspapers and magazines. Name some skills that a student must have to read and/or construct graphs, tables, and charts.

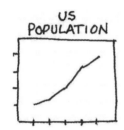

COMPREHENSION
Answer this question: When are charts, tables, and graphs most helpful as a source of information to the reader or researcher?

APPLICATION
Survey the students in your English or language arts class on one of these topics: (1) How much time do you spend weekly on your homework in this class? (2) What are your favorite books or who are your favorite authors? (3) What types of assignments do you enjoy most/least in this class? (4) What topics that we study in this class do you find most/least interesting and why? Then, construct a chart, table, or graph to show the results of your survey.

ANALYSIS
How are charts, tables, and graphs alike and how are they different?

SYNTHESIS
Relate each of these statements to your study of charts, tables and graphs. What do they mean or imply as figures of speech? (1) Be sure to chart your course in school if you want to be successful! (2) We will table these arguments until our next meeting. (3) Don't telegraph your intentions!

EVALUATION
Agree or disagree with this statement: A chart, table, or graph is worth a thousand words!

Ignite Student Intellect and Imagination in Language Arts, by Sandra L. Schurr and Kathy L. LaMorte. Published by National Middle School Association, 2007.

PREPARATION OF STUDY NOTES FOR A TEST

KNOWLEDGE
Record key terms and their definitions on a series of note cards for a subject you are studying and that you will be tested on later. Put one term/definition per card.

COMPREHENSION
Prepare a set of cards that include the important ideas and their supporting details from all assigned readings and lectures in class that are related to your topic. Put one major concept with its details per card.

APPLICATION
Construct a number of detailed graphic organizers on your subject as study aids making certain there is only one completed organizer on a card.

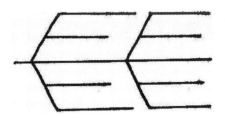

ANALYSIS
Review all of your cards and information-----key terms/definitions, important ideas/details, and relevant graphic organizers. Group the cards by topics and subtopics related to your subject. Paperclip those that belong together.

SYNTHESIS
Use a highlighter to identify key words or phrases on each of the cards as a summary or memory aid. Use the margins of the cards to record book pages/sources in case you need to go back and review the material on your subject. Try writing out a set of hypothetical questions that the teacher might ask you about your subject/topic and then try answering them without the use of your note cards.

EVALUATION
Can you think of a better way to prepare for a test than through the use of note cards to study? Defend your position.

Ignite Student Intellect and Imagination in Language Arts, by Sandra L. Schurr and Kathy L. LaMorte. Published by National Middle School Association, 2007.

RESEARCH INFORMATION SOURCES

KNOWLEDGE

Write down a list of print and non-print resources that one might use when conducting research on assigned topics or issues. Circle those that are available electronically online.

COMPREHENSION

Briefly explain when a student might use each of these print and non-print resources as a tool for research: Abridged Readers' Guide to Periodical Literature, Almanac, Atlas, Bartlett's Familiar Quotations, Guinness Book of World Records, Library Computer Catalog, Periodical Index, Webster's Biographical Dictionary, Webster's New Geographical Dictionary, and World Almanac.

APPLICATION

Practice using the interview process as a tool for gathering input on a research assignment. Select a person to interview and determine whether it is to be a phone interview, a personal one-on-one interview, or an online interview. Prepare a set of questions to ask the interviewee. Conduct the interview and compile the results.

ANALYSIS

Databases and computer networks are also excellent sources of information for conducting research on a topic. Point out some of the advantages and disadvantages of using electronic resources for information when compared to print and nonprint resources.

SYNTHESIS

Artifacts and personal letters/journal entries can also be used as tools for conducting research. Give some creative examples of when someone might want to examine artifacts/letters/journals as a basis for their investigative work.

EVALUATION

Is it easier to validate the authenticity of information gathered for research from the Internet or from print and nonprint sources housed in a school/community library? Give reasons for your answer.

Ignite Student Intellect and Imagination in Language Arts, by Sandra L. Schurr and Kathy L. LaMorte. Published by National Middle School Association, 2007.

WRITING THE CLASSROOM REPORT OR RESEARCH PAPER

KNOWLEDGE

Why do you think teachers ask you to write reports or papers on a subject in class? How do you feel about these report assignments? Try putting these steps of writing a report/paper in the correct order: Gathering the details; Recording your information; Choosing a subject; Writing the first draft; Making corrections; Giving credit for sources used; Outlining your information; Organizing your information; Adding a bibliography.

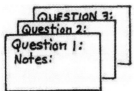

COMPREHENSION

Choose a report topic and develop a set of questions to help you organize your reading, research, and note taking. Put each question on a separate note card.

APPLICATION

Conduct research on your topic and answer the questions on your note cards. Be sure to record the source of your information on each card as well.

ANALYSIS

Review each of your note cards and select one of the cards to be the main point of your report. Then, arrange the rest of your note cards in the best possible order. Next, use the cards to write a comprehensive outline for your report. Finally, write your report based on the outline. Be sure to include an introductory paragraph, a concluding paragraph, and a series of related paragraphs between the two as the body of the report.

SYNTHESIS

Think of creative ways to "jazz up" your report. Try typing your report on the computer and using color, graphics, varied print fonts, and unusual spacing formats to add interest for the reader.

EVALUATION

Review your final report and look for an interesting opening line, a statement of the specific report topic, several strong points (one per paragraph) with supporting details, a strong conclusion, and notes giving credit for information.

Ignite Student Intellect and Imagination in Language Arts, by Sandra L. Schurr and Kathy L. LaMorte. Published by National Middle School Association, 2007.

TROUBLESHOOTING POTENTIAL RESEARCH/REPORT PROBLEMS BEFORE THEY BEGIN

KNOWLEDGE
Begin looking for a research topic by looking at yourself and your interests. Circle any of the following topic ideas that appeal to you:

(1) Current events;
(2) Nature;
(3) Environmental issues
(4) Social issues;
(5) Political issues;
(6) Economic issues;
(7) Interesting Places;
(8) Theories;
(9) Famous people;
(10) Controversial issues.

COMPREHENSION
It is important to provide your reader audience with information that is new and interesting. Think of at least three questions you might use to evaluate an audience for your research topic.

APPLICATION
To write a good research paper, you must find good information. Determine how you would evaluate a source as being appropriate for your research paper and discuss reasons for identifying these sources correctly. How does one avoid "plagiarism" as part of this process?

ANALYSIS
You can take notes from a source in one of three ways: direct quotation, paraphrase, or summary. What are advantages and disadvantages of each?

SYNTHESIS
Propose several ideas for making one's research paper more creative and interesting. Consider everything from ways to grab the reader's attention to assembling and getting the information down on paper.

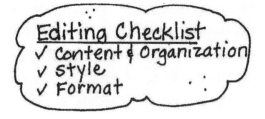

EVALUATION
One should edit a research paper to make it as correct as possible. Develop an editing and mechanics checklist to use for this purpose.

Ignite Student Intellect and Imagination in Language Arts, by Sandra L. Schurr and Kathy L. LaMorte. Published by National Middle School Association, 2007.

STAGING A DEBATE FOR THE AUDIENCE

KNOWLEDGE
Define "debate" and state its overall purpose.

COMPREHENSION
Give examples of famous debaters and debates from history.

APPLICATION
Organize a debate between you and some of your classmates. Choose a resolution topic, conduct research on the topic, and develop a series of affirmative and negative arguments related to the topic. Determine which student participants will support the affirmative position and which will support the negative position.

ANALYSIS
Point out the challenges and roadblocks which good debaters must learn to deal with.

SYNTHESIS
Use the following set of rules for a 25 minute debate on your topic:
1. (3 minutes) First Speaker for the affirmative makes opening statement.
2. (3 minutes) First Speaker for the negative makes opening statement. (No questions or interruptions allowed during these times.)
3. (3 minutes) Second and Third Speakers for the affirmative stand and are questioned by the Second and Third Speakers for the negative. (They answer as best they can.)
4. (3 minutes) Second and Third Speakers reverse roles. The affirmative side questions, and the negative side answers.
5. (5 minute) Recess for both sides to discuss closing ideas with final speakers.
6. (2 minutes) Fourth Speaker for the affirmative side summarizes.
7. (2 minutes) Fourth Speaker for the negative side summarizes
8. (4 minutes) Judges withdraw to determine the winner of the debate.

EVALUATION
Develop a rubric for the judges to use in selecting the winning side of your debate. What criteria should they use in making their decision?

Ignite Student Intellect and Imagination in Language Arts, by Sandra L. Schurr and Kathy L. LaMorte. Published by National Middle School Association, 2007.

SECTION H

Bloom Sheets related to the standard of
Using a Variety of Resources to Gather and Synthesize Information

Standard 8

Students use a variety of technological and information resources (e.g., libraries, databases, computer networks, video) to gather and synthesize information and to create and communicate knowledge.

YOUR COMPUTER AS INFORMATION TOOL

KNOWLEDGE
Name the three main parts of a computer system.

COMPREHENSION
Summarize the importance of these Internet basics and their purposes: World Wide Web (WWW or web), Uniform Resource Locators (URLs), Browsers, Home/Web Pages, and Search Engines.

www.google.com

APPLICATION
Demonstrate how one goes about finding reliable information on the Internet by preparing an outline, flow chart, or detailed description of a research-based homework assignment that you have recently completed (or plan to complete) for a class.

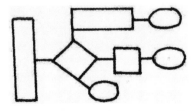

ANALYSIS
Infer when you are more likely to use the Internet as an information source for an assigned research task than you are the Dewey Decimal System and vice versa.

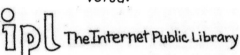
The Internet Public Library

SYNTHESIS
Create a glossary of Internet sites that you have found to be reliable sources of information for school-related research assignments. Some good questions to ask about a Web site might be:
(1) Who are the sponsors and are they an established group?
(2) Is there useful information available and are there documented sources for the information cited?
(3) Are the authors of the information qualified and how do you know?
(4) Is the material and site up to date and is it easy to review and understand?

Bartleby.com

EVALUATION
React to this statement based on your own experiences: Computers have greatly enhanced the organization and use of library references and resources. Be specific in your comments.

Ignite Student Intellect and Imagination in Language Arts, by Sandra L. Schurr and Kathy L. LaMorte. Published by National Middle School Association, 2007.

BOUNDARIES AND BARRIERS OF COMPUTERS

KNOWLEDGE

Write down at least twenty ways that computers have changed our lives. Here are two to get you started:

(1) They fly planes;
(2) They give you money at a bank.

COMPREHENSION

Explain how you think authors use computers to write, edit, and rewrite entire books.

APPLICATION

Investigate the many ways you can use computers to help you complete a major research assignment. Develop a list of the computer assisted tools and techniques which you can access to help you write a quality research paper from beginning to end.

ANALYSIS

It is likely that a student of the future will soon find today's library of print fiction/nonfiction books and reference resources out-of-date. How would you like to find yourself limited to only electronic books and research resources for locating information without the option of using print materials as well? Are there advantages to both? If so, what are they?

SYNTHESIS

Write a position paper that discusses several of the legal and ethical issues that one now encounters in cyberspace.

EVALUATION

Recommend some student guidelines that are designed to create boundaries and barriers to promote safety on the Internet.

Ignite Student Intellect and Imagination in Language Arts, by Sandra L. Schurr and Kathy L. LaMorte. Published by National Middle School Association, 2007.

LEARNING MORE ABOUT MEDIA AND TECHNOLOGY AS A TOOL IN THE LANGUAGE ARTS CLASSROOM

KNOWLEDGE

Recall ways that you have used media and technology in your language arts classroom as a learning tool so far this year.

COMPREHENSION

Describe ways that media and technology can enhance one's creativity in the traditional classroom.

APPLICATION

Use a piece of software to write a story, compose a poem, or produce a play.

ANALYSIS

Discover a variety of effective Web sites that kids your age could visit for help with their writing and research assignments on the Internet. Point out things to look for when visiting each site.

SYNTHESIS

Design a kid's page for the Internet that focuses on interesting and unusual topics or springboards for journal entries, research papers, and creative writing topics.

EVALUATION

Determine why educational videos, CD's, and other visual aids can be more effective tools for teaching new concepts than print materials for many students.

Ignite Student Intellect and Imagination in Language Arts, by Sandra L. Schurr and Kathy L. LaMorte. Published by National Middle School Association, 2007.

TECHNOLOGY CHANGES AND CHALLENGES

KNOWLEDGE
List and define ten important terms or concepts related to media and technology.

COMPREHENSION
Prepare five possible questions for the answer: "Internet."

APPLICATION
Organize and moderate a panel discussion to help others better understand the multiple and exciting uses of the computer as an instructional tool in today's classroom.

ANALYSIS
Stage a debate about the pros and cons of relying on electronic spelling and grammar aids for your writing assignments in lieu of developing your own competencies in these areas.

SYNTHESIS
Write a brief "how to" explanation for using a piece of school-related computer software with which you are familiar.

EVALUATION
Research to determine what types of writing-related software programs or websites are the best learning tools for kids your age. Establish a rating scale for judging these options based on your findings.

Ignite Student Intellect and Imagination in Language Arts, by Sandra L. Schurr and Kathy L. LaMorte. Published by National Middle School Association, 2007.

USING WRITING PROMPTS TO EXAMINE MEDIA AND TECHNOLOGY MYTHS

KNOWLEDGE

Use the dictionary to define the concept of "myth." Then complete these writing tasks which challenge some modern day myths related to media and technology in our present day schools and everyday lives.

COMPREHENSION

Give examples of why this observation isn't true: "Teachers resist learning and using new technologies in the classroom because they are intimidated by their power and by their potential to make teachers less important in the learning process."

APPLICATION

Relate to this statement: "Because we live in a digitized world, we are losing our 'humanness' and our need for human touch."

ANALYSIS

Criticize this belief: "Electronic and Internet games inhibit the development of interactive communication skills between and among kids of this generation."

SYNTHESIS

Formulate a response to this myth: "Technology reduces learning and knowledge to information that is limited to hard-core facts, data-bases, and statistics eliminating opinions, feelings, and personal thoughts."

EVALUATION

Argue against this idea: "As people of all ages depend more and more on video, DVD, radio, television, and film for education and entertainment, book sales will continually decrease and consequently so will world literacy."

Ignite Student Intellect and Imagination in Language Arts, by Sandra L. Schurr and Kathy L. LaMorte. Published by National Middle School Association, 2007.

INSTANT MESSAGING

KNOWLEDGE
Define "text or instant messaging" as it relates to e-mailing and computer lingo.

COMPREHENSION
Explain the origins and popularity of instant messaging as a means of communication among teens today.

APPLICATION
Demonstrate your IM (Instant Messaging) IQ by translating these terms: ATM, AND, AI, AIS, BFA, BBL, CNG, COO, FAAK, GAC, H, HIG, LOLA, J/K, E2EG, MOS, PLOS, POS, TAW, BIW, TSR, SBB, WB, ROFL, and SIT.

ANALYSIS
Examine these adult forms (acronyms) of instant messaging and see how well you do: AARP, AMA, FCC, DOT, DNC, RNC, FDR, JFK, CFO, CEO, YTD, PGA, LPGA, GNP, MGM, SNL, UPC, HMO, CNN, MLS, FEMA, WASP, B&B, RSVP, and BBC.

SYNTHESIS
Compose an original short story, skit, or dialogue that uses both teenage and adult examples of instant messaging to show how they are used in two different worlds.

EVALUATION
It has been observed that both overuse of e-mail and instant messaging among young people today have impacted negatively on their ability to write, speak, and spell words correctly in school. What do you have to say about this?

Ignite Student Intellect and Imagination in Language Arts, by Sandra L. Schurr and Kathy L. LaMorte. Published by National Middle School Association, 2007.

ONLINE DICTIONARY AND THESAURUS USAGE

KNOWLEDGE
Look at a typical page of a dictionary you find online. List all of the uses for a dictionary besides looking up the spelling or meaning of a word.

COMPREHENSION
Choose a favorite word of yours and prepare a king-size entry for your word that mimics a typical entry in the online dictionary.

APPLICATION
Construct a treasure hunt or test instrument that would challenge someone's ability to use the online dictionary effectively.

ANALYSIS
Look at a typical page of a thesaurus you find online. How does it differ from a dictionary entry?

SYNTHESIS
Choose a subject you are studying in a science, math, or social studies class at the present time. Create a mini-dictionary of its important words, terms, or concepts and store it online.

EVALUATION
When engaged in a creative writing activity or homework assignment, are you more likely to make good use of an online dictionary or a standard print dictionary as found in the home or school. Give reasons for your answer.

Ignite Student Intellect and Imagination in Language Arts, by Sandra L. Schurr and Kathy L. LaMorte. Published by National Middle School Association, 2007.

SECTION I

Bloom Sheets related to the standards of
Understanding Diversity and Using a Non-English First Language

Standards 9 & 10

Students develop an understanding of and respect for diversity in language use, patterns, and dialects across cultures, ethnic groups, geographic regions, and social roles.

Students whose first language is not English make use of their first language to develop competency in the English language arts and to develop understanding of content across the curriculum.

DIVERSITY IN LANGUAGE USE AND DIALECTS

KNOWLEDGE
Answer this question: What is a dialect and how do you recognize one?

COMPREHENSION
In the United States, various regions have different dialects. Give some examples of these American dialects.

APPLICATION
Discover what these words mean in their designated regions:
New England: *grinder, cabinet, show, and yardman*
Southern Mountains of United States: *man-child, kinfolks, treads, and wax*
Hawaii: *bla, bedclothes, package, broke, and lab*

ANALYSIS
Infer what this regional expression means: "Don't 'sweet talk' your parents!"

SYNTHESIS
Make up a comic strip or cartoon series that features characters from a particular part of the country who speak their own language pattern or dialect.

EVALUATION
Children with special dialects who move from one part of the country to another sometimes have trouble with diction and spelling in school. Assess why this is true!

Ignite Student Intellect and Imagination in Language Arts, by Sandra L. Schurr and Kathy L. LaMorte. Published by National Middle School Association, 2007.

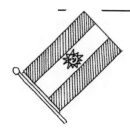
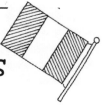

MULTICULTURAL MEANS MULTI-LANGUAGES

KNOWLEDGE
Many students in our schools throughout the United States are designated as ESOL students. What does this acronym stand for?

COMPREHENSION
It is not uncommon to have multiple languages spoken in many schools across the country. For example, one Florida school has these languages present among the student body: Armenian, Arabic, Bulgarian, Czech, French, Greek, Haitian-Creole, Hungarian, Mandarin, Polish, Portuguese, Russian, Serbian, Slovak, Spanish, and Ukrainian. Point out the problems that some of these students will have when taking a typical language arts or English course in their elementary, middle, and high school years.

APPLICATION
Limited English-speaking students have the right to a quality education in American schools. Demonstrate one or more strategies that teachers use to help these students acquire competency in the writing and speaking of English.

ANALYSIS
Compare and contrast the word/spelling/pronunciation of each of these common phrases among the following European languages: English, French, German, Italian, and Spanish. Hello or good day; Thank you; Excuse or pardon me; Yes; No; Goodbye/so long.

Bon jour   Buon giorno

Buenos días   Guten tag

SYNTHESIS
Plan a Unity and Nationality Day for your class or school. Encourage all students to wear a piece of clothing recognizing their heritage and to teach one another how to say specific words or phrases in their language throughout the day.

EVALUATION
Decide what would be most difficult for you if you moved to a different country and did not speak the language.

HOW DOES LANGUAGE VARY WITH SOCIAL ROLES?

KNOWLEDGE

When you want to point out that something is painfully or embarrassingly obvious, you say:
A. Ouch!   B. Duh!   C. Ping!
D. Been there, done that!
(Answer: B)

COMPREHENSION

Explain how your communication skills (choice of words, voice inflections, attitudes) differ with varied social roles or groups. For example, do you converse with your parents the same way you do with your friends? Do you converse with your teachers the same way you do with your peer group?

APPLICATION

It has been said that "a worthy catch phrase packs a little oomph."
What are some "catch phrases" that you use with your friends? With family members?

ANALYSIS

Catch phrases have unique staying power. Unlike slang or jargon, which often disappear from overuse, these phrases (perceived as clever and ironic, heavy on voice inflection and attitude) are part of our everyday social discourse. Determine situations or times when you might say each of these catchy phrases to a friend or family member:
(1) Yeah, right;
(2) Don't go there;
(3) Hel-lo;
(4) I don't think so!
(5) Phh-leeze;
(6) Get over it!
(7) I'm not having this conversation.

SYNTHESIS

Create a humorous skit that focuses on the theme: "A worthy catch phrase packs a little oomph." Use as many of these expressions as you can in the dialogue. Here are two more to get you thinking . . . "Whatever," "Slam dunk."

EVALUATION

Some language experts feel that "catch phrases" can be rude, can stilt conversations, and can stunt our rhetorical prowess. What do you think? Support your answer.

Ignite Student Intellect and Imagination in Language Arts, by Sandra L. Schurr and Kathy L. LaMorte. Published by National Middle School Association, 2007.

AMERICAN ENGLISH AND BRITISH ENGLISH

KNOWLEDGE
Locate Great Britain on a world map. Note its proximity (or lack of it) to the United States.

COMPREHENSION
It has been said that the United States and Great Britain are two nations divided by a single language. Explain why this is true.

APPLICATION
Spellings of the same words can be decidedly different in the United States and Great Britain. Two examples of the variances between American and British spellings are: check & cheque; color & colour; theater & theatre. Collect pairs of other words where this is true and write them down.

ANALYSIS
The two versions of the English language also differ when it comes to the names of many everyday objects and events. Analyze these British terms and come up with the American equivalent to them: water closet (WC), biscuit, nappy, draughts, tap, petrol, flat, cooker, braces, lorry, boot, tube, vest, holiday, and waistcoat.

SYNTHESIS
Construct a sentence that shows the different use of quotation marks in a sentence between the two countries.

EVALUATION
Do you think the British dialect is different from that of the United States? Explain.

Ignite Student Intellect and Imagination in Language Arts, by Sandra L. Schurr and Kathy L. LaMorte. Published by National Middle School Association, 2007.

COLORFUL LANGUAGE EXPRESSIONS

KNOWLEDGE

Expressions make language exciting because they color our thoughts and ideas. Expressions are words or phrases that are not found in the dictionary, but rather are handed down through generations and cultures via historical events, ethnic myths or legends, and people over time. What are some personal expressions that you use on a daily basis to convey your feelings or thoughts?

COMPRHENSION

Adages, proverbs, and maxims are interesting because they tell truths about life or human nature. In your own words, explain what you think this expression means: "Don't put all your eggs in one basket."

APPLICATION

Idioms are expressions that can't be understood literally from the individual words contained in them. Tell about a time when you were "in the dog house," were "mad as a hatter," or the "apple of one's eye."

ANALYSIS

Some expressions are taken from literature such as Greek mythology or Aesop's fables. Research to find out the origin of these two literary expressions: "Achilles' heel" and "Cry wolf."

SYNTHESIS

Clichés are tired expressions that have been overused when making a strong impression and should be avoided in speaking and writing. Use the Internet to locate some clichés and create a series of drawings to show how ridiculous they can be when interpreted literally rather than figuratively.

EVALUATION

Survey the students in your class to compile a list of their favorite teenage expressions.

Ignite Student Intellect and Imagination in Language Arts, by Sandra L. Schurr and Kathy L. LaMorte. Published by National Middle School Association, 2007.

FUN WITH PROVERBS

KNOWLEDGE
Record a brief definition/ description of a proverb and write down several examples. Try to include proverbs representative of different cultures or ethnic groups.

 a bird in the hand...

COMPREHENSION
Rewrite your proverbs in hieroglyphics.

APPLICATION
Express your proverbs using sophisticated synonyms for all key words in each saying. Notice how different these proverbs now appear in both content and appearance.

ANALYSIS
Draw conclusions as to the when and why or origin of each proverb. What common activities or everyday experiences might have prompted these "wise sayings"?

 a penny saved...

SYNTHESIS
Create an original story that uses one of the proverbs as its topic sentence or concluding statement.

EVALUATION
Judge the truth of this statement: Proverbs provide generations with a means of preserving one's culture. Defend your position.

Ignite Student Intellect and Imagination in Language Arts, by Sandra L. Schurr and Kathy L. LaMorte. Published by National Middle School Association, 2007.

SIGN AND SYMBOL LANGUAGES

TÒΦῶS 吉国川9岩

KNOWLEDGE

Use each letter of the alphabet to write down an English word or phrase that tells something about yourself.

COMPREHENSION

Explain the purpose and use of sign language in our world today. Show examples of letters and words that are made by shaping the fingers and hands in different ways.

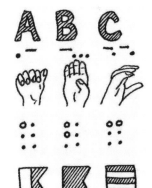

APPLICATION

The Braille alphabet consists of raised dots that are used to represent each letter in the alphabet. Locate information about the Braille system and the dot formations that are used for each letter of the alphabet. Send a message to someone using this system of communication.

ANALYSIS

Compare the Braille alphabet with that of the Morse Code. How are they alike and how are they different? Compare various letters and words.

SYNTHESIS

Sailors use the semaphore to communicate with one another at sea. In semaphore, brightly colored flags in different patterns and angles are used to represent different letters and alphabets. Research more about the semaphore alphabet. Construct a set of flags and practice sending messages back and forth to a friend.

EVALUATION

Use the Internet to learn something about other alphabets like those used by the Chinese, Japanese, Greeks, or Egyptians. Judge their difficulty as compared to that of the English. Which would be harder for a foreigner to learn and why?

Ignite Student Intellect and Imagination in Language Arts, by Sandra L. Schurr and Kathy L. LaMorte. Published by National Middle School Association, 2007.

SECTION J

Bloom Sheets related to the standard of
Participation in a Variety of Literacy Communities

Standard 11

Students participate as knowledgeable, reflective, creative, and critical members of a variety of literacy communities.

EXTRA! EXTRA! READ ALL ABOUT IT

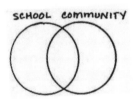

KNOWLEDGE

Locate a copy of your community newspaper. Use the index on the front page of the newspaper and locate the various sections listed. Browse through these sections to familiarize yourself with their content.

COMPREHENSION

News and feature stories use a special structure called a pyramid to report their factual and/or human interest information in an article. In your own words, explain what is meant by the "who, what, when, where, why, and how" of a newspaper story.

most newsworthy

least newsworthy

APPLICATION

Create a mini-newspaper on a topic of your choice. You might want to write about yourself and your family or an interesting math/science/social studies topic currently being studied in class. Try to construct your mini-newspaper report so that it includes the following components: Banner Headline, News Story, Feature Story, Letter to the Editor or Editorial Cartoon, Comic Strip, Classified Ads, Display Ad, and Book/Movie/Play Review.

ANALYSIS

Obtain a copy of your classroom or school newspaper and compare and contrast it with a copy of your community newspaper. Note the similarities and differences.

SCHOOL   COMMUNITY

SYNTHESIS

Think of a creative and catchy name for your mini-newspaper project that will attract readers of all types.

EVALUATION

It has been said that newspapers in the United States are decreasing in circulation numbers because young adults today, unlike older generations, are not subscribing to their local paper. They prefer to download news from the Internet or listen to news broadcasts from the media. Why do you think this is the case?

Ignite Student Intellect and Imagination in Language Arts, by Sandra L. Schurr and Kathy L. LaMorte. Published by National Middle School Association, 2007.

MAGAZINES COME IN ALL SHAPES AND SIZES

KNOWLEDGE
Make a list of magazines that you like to read when they are available to you at home, at school, in a barber/beauty salon, doctor/dentist office, library, reception area, or in a retail outlet.

COMPREHENSION
There are magazines published for every special interest group. Suggest a specific magazine title for at least 20 different periodicals that cater to varied markets such as sports, health/fitness, news, teenagers, entertainment, hobbies, food, family/parenting, computers, etc.

APPLICATION
Examine several different magazines from the Knowledge and Comprehension Levels and list the features that are common to most magazines regardless of their focus.

ANALYSIS
Point out the number and types of advertising that permeate every magazine on the market. What percentage of the pages in your favorite magazine are devoted to advertising copy?

SYNTHESIS
Write a letter to a magazine of your choice commenting on one or more articles that had special appeal to you.

EVALUATION
Determine what skills and experiences it would take to be a major author or contributor to your favorite magazine!

Ignite Student Intellect and Imagination in Language Arts, by Sandra L. Schurr and Kathy L. LaMorte. Published by National Middle School Association, 2007.

HOW TO READ A HOW TO BOOK OR MANUAL

KNOWLEDGE

Recall a time when you had to read and follow directions carefully to complete an important task or function.

COMPREHENSION

Review the pages of any manual or set of directions you have for operating an electronic device in your home such as computer game, cell phone, I-Pod, or installing a piece of software. Explain what special skills it takes on your part to decipher the information.

APPLICATION

Discover how illustrations and diagrams are often used to accompany explanations or directions in most "how-to" manuals, books, or printed directions.

ANALYSIS

Debate this position: The decoding and comprehension skills required to read and follow the plot line of a novel are much different from those required to read and follow a set of directions or guidelines for putting something together or figuring something out.

SYNTHESIS

Think of a complex task that you would like someone to do. Then write out a complete set of directions so that person can complete the task to your satisfaction. How difficult is this to do?

EVALUATION

Would you rather spend an afternoon reading for pleasure or reading to learn how to do something? Support your position.

Ignite Student Intellect and Imagination in Language Arts, by Sandra L. Schurr and Kathy L. LaMorte. Published by National Middle School Association, 2007.

THE GIFT OF A LETTER

KNOWLEDGE

Briefly identify the basic parts of a friendly letter and of a business letter.

COMPREHENSION

Give two or three examples of times when you or someone else might choose to write a friendly letter.

APPLICATION

There are three basic types of business letters that you might want to write as a teenager. They are: Letter of Inquiry in which you ask for information or answers to your questions; Letter to an Editor or

Official in which you want to complain, compliment, or clarify; Letter of Complaint in which you want to discuss a problem and any action required to solve the problem. Write a business letter of your own based on one of these situations.

ANALYSIS

Discuss the advantages and disadvantages of writing and sending a letter over a telephone call or an e-mail to a person, business, or institution.

SYNTHESIS

Create your own stationery, voice mail message, and/or e-mail address. Use them when you need them!

EVALUATION

Write your own set of behavioral Do's and Don'ts for writing personal letters, making personal phone calls, and sending personal e-mail messages.

Ignite Student Intellect and Imagination in Language Arts, by Sandra L. Schurr and Kathy L. LaMorte. Published by National Middle School Association, 2007.

JOURNAL WRITING MADE INTERESTING

KNOWLEDGE

Why do you think many teachers in today's schools require journal writing of their students? What are the benefits to both teachers and students alike? What kinds of prompts or guidelines for entries do you prefer?

COMPREHENSION

One type of prompt for a journal entry is a famous quotation. Give

some examples
of quotations
that you like
and react to each of them.

APPLICATION

Another type of prompt for a journal entry is a comic strip or editorial cartoon. Collect some interesting examples and discuss each one's purpose and position on an issue or situation.

ANALYSIS

A popular prompt for a journal entry can also be a poem or an excerpt from a popular novel or short story. Select some type of literary writing that you enjoy and point out some interesting prompts that a teacher might assign as a springboard for journaling.

SYNTHESIS

Portraits, famous masterpieces, or even colorful magazine illustrations/photographs can be used as writing prompts for students. Assume the role of the teacher and choose an artistic picture to show in class. Make up several different prompts for kids to use in writing their perceptions and interpretations of the art item.

EVALUATION

Grading or assessing the quality of a response to a writing prompt provides a challenge for both student and teacher alike. Recommend a set of guidelines for the teacher to consider when judging the overall value of student responses to a given writing prompt.

Ignite Student Intellect and Imagination in Language Arts, by Sandra L. Schurr and Kathy L. LaMorte. Published by National Middle School Association, 2007.

READING QUALITY LITERATURE

KNOWLEDGE

There are several established measures for identifying quality literature in our world today. Identify each of these measures for children's books: Newbery Medal and Caldecott Medal. List five titles of books that fall into each category.

COMPREHENSION

Middle school and high school teachers often require their students to read several of the "classics." Explain what types of books would be considered a classic and give several examples.

APPLICATION

Adult reading also has its awards for good reading and writing. They range from Book-of-the-Month Club and Oprah Book Club award-winning choices to the famous Pulitzer Prize and Nobel Prize for Literature. Use the Internet to find out more information about these awards and the books chosen to receive them.

ANALYSIS

Many school districts have developed core or suggested reading lists for each grade, kindergarten through grade twelve. These books represent the best in literature from several different perspectives. Pretend you have been chosen to prepare a set of guidelines for educators to use when selecting books for this list. What guidelines or criteria should be used in this process? What are five books you would recommend that meet your criteria for your age group?

SYNTHESIS

Create a new BOOK AWARD to be given to authors/books that write specifically for kids your age. Design and describe the award. Who would you give it to this year?

EVALUATION

What is the best book that you have ever read? What criteria did you use to make this choice?

Ignite Student Intellect and Imagination in Language Arts, by Sandra L. Schurr and Kathy L. LaMorte. Published by National Middle School Association, 2007.

SECTION K

Bloom Sheets related to the standard of
Enabling Students to Use Language to Enhance Their Own Purposes

Standard 12

Students use spoken, written, and visual language to accomplish their own purposes (e.g., for learning enjoyment, persuasion, and the exchange of information).

WRITING TO PERSUADE

KNOWLEDGE
Distinguish between fact and opinion. Select a topic you know something about. State a fact about the topic as well as an opinion you have on the topic.

COMPREHENSION
In your own words, explain the difference between fact and opinion.

APPLICATION
Choose something that you feel strongly about----an opinion you have which can be supported by documented facts. Then, write an opinion (topic) statement followed by a series of proven facts. Organize your facts in such a way that they lead to a logical and believable conclusion.

ANALYSIS
Go back over your written work and look for these problem areas:
(1) Did you include statements that jump to a conclusion;
(2) Did you include statements that contain a weak or misleading comparison;
(3) Did you avoid statements that exaggerate the facts or mislead the reader;
(4) Did you include statements that appeal only to the reader's feelings without facts to back them up?
(5) Did you include statements that contain part of the truth, but not the whole truth? Correct any statements that are not meaningful or well thought out.

SYNTHESIS
Be creative and think up a clever title for your piece of persuasive writing. How did you grab the reader's attention? How did you convince the reader of your opinions?

EVALUATION
Give yourself a letter grade on the writing. What criteria did you use to assess your work? Be specific in your comments.

Ignite Student Intellect and Imagination in Language Arts, by Sandra L. Schurr and Kathy L. LaMorte. Published by National Middle School Association, 2007.

GETTING THE MOST OUT OF PERSONAL WRITING

KNOWLEDGE

Record some types of personal writing that you or family members might do at home throughout the week, month, or year. These may be hand-written or electronically-recorded.

To Do...
1.
2.
3.
4.
5.

COMPREHENSION

There are five different journal writing types. A Dialogue Journal involves an ongoing conversation between you and another person who share a common interest or need. A Diary is a record of your daily events, experiences, or observations.

Dear Diary,

A Learning Log helps someone like you to examine personal thoughts, feelings, and reactions to ongoing course work. A Specialized Journal focuses on a single event or experience of importance to you. A Travel Log is a record of your vacationing or traveling memories. Describe a time or situation when you actually engaged in one or more of these five journal writing options.

APPLICATION

Experiment with each of the personal writing tasks described at the Comprehension Level during the semester and determine how each one works for you over time.

ANALYSIS

Journal writing benefits you in many ways. It can provide you with a valuable record of your ideas, thoughts, memories, and experiences. It offers you an interesting way to practice writing. It helps you create new ideas and offers you a format for recording original or novel ideas. It serves as a resource for recording facts, opinions, and details for future assignments/reflections. Which of these benefits do you think will serve you best and why?

SYNTHESIS

Use the computer to create your own journal format for personal use at home or at school.

EVALUATION

Do you prefer recording your journal entries as an electronic file or as a hand-written entry in a paper/pencil diary format? Explain.

Ignite Student Intellect and Imagination in Language Arts, by Sandra L. Schurr and Kathy L. LaMorte. Published by National Middle School Association, 2007.

GREAT WAYS TO SHARE A BOOK

KNOWLEDGE
Record the title, author, publisher, and copyright date of a novel that you are going to read and share as a formal book report assignment.

COMPREHENSION
Describe a special situation or event in the story that caused you to feel any five of these ten emotions: Fear, Disgust, Sadness, Pity, Joy, Hopelessness, Anger, Frustration, Hope, or Worry.

APPLICATION
Construct a story grid or graph for your novel. Write the story's ten main events along the bottom

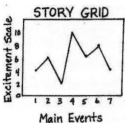

of the grid and an excitement rating scale of 0 through 10 for each event along the left-hand side of the grid. Plot the excitement rating for each event and connect the points to complete your graph.

ANALYSIS
Choose one of the important characters from the story. Analyze his/her character traits and behaviors/attitudes through-out the story. Then write a letter giving some good advice to this character.

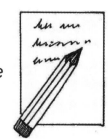

SYNTHESIS
Pretend this novel was going to be made into a movie. Write down the actors/actresses you would cast for the protagonist and the antagonist and give reasons for your choices.

EVALUATION
Decide on three good questions that you would want to ask if you were given an opportunity to interview the author of this novel.

Ignite Student Intellect and Imagination in Language Arts, by Sandra L. Schurr and Kathy L. LaMorte. Published by National Middle School Association, 2007.

HOW TO DESIGN AND PUBLISH GREETING CARDS

KNOWLEDGE

Think of all the times during the year when one might send out a greeting card to an individual. Visit a Hallmark store and note the many different and unusual occasions for sending these messages----new job, loss of pet, divorce, or friendship---- to name a few.

COMPREHENSION

Investigate the availability of sending online or electronic greeting cards. Prepare a summary of your information so that others can benefit from your research.

APPLICATION

Show others that you "care enough to send the very best" by creating and sending a greeting card to a friend, a family member, a celebrity you admire, or even a fictional character in a book.

ANALYSIS

Examine several Web sites for making and sending greeting cards including pop-up cards. Collect and store famous quotes and messages for these potential cards.

SYNTHESIS

Think of ways that greeting cards could be used as part of the curriculum for teaching poetry. When you create greeting cards, you are practicing the skills of rhyme, surprise, humor, and conciseness.

EVALUATION

Assess the possibility of creating a card store in your classroom to meet the greeting card needs of your school or to raise funds for a special field trip or project.

Ignite Student Intellect and Imagination in Language Arts, by Sandra L. Schurr and Kathy L. LaMorte. Published by National Middle School Association, 2007.

WRITING TO SPEAK OR SPEAKING TO WRITE

KNOWLEDGE

Locate a famous speech from history and practice reading it aloud. Lincoln's "Gettysburg Address," Martin Luther King's "I Have A Dream" speech, or Julius Caesar's "Friends, Romans, Countrymen" text are all good examples.

COMPREHENSION

Invite a speech writer to your class and have him/her discuss how to go about writing and giving a speech. Ministers, lawyers, staff writers for public relations firms, or members of the local "Toastmaster's Club" are all options to consider. Prepare a letter of invitation to this individual explaining the purpose for the visit.

APPLICATION

Organize a group of students to write and deliver a series of morning announcements for the beginning of each class period making them both informational and inspirational in their message.

ANALYSIS

Determine which of these speech formats has the most appeal to you and why:
(1) All talk;
(2) Talk and show; or
(3) Talk and audience participation.

SYNTHESIS

Prepare your speech by following these guidelines:
(1) Pick your subject;
(2) Establish your purpose;
(3) Identify your audience;
(4) Know your time limits;
(5) Decide on your speech format;
(6) Research your subject;
(7) Outline your main ideas;
(8) Draft your text;
(9) Rehearse your speech;
(10) Analyze your audience's reaction.

EVALUATION

Construct a rubric that establishes a rating scale and criteria for judging the quality of your speech. How did you do?

Ignite Student Intellect and Imagination in Language Arts, by Sandra L. Schurr and Kathy L. LaMorte. Published by National Middle School Association, 2007.

A GOOD SPEECH IS LITERATURE TOO!

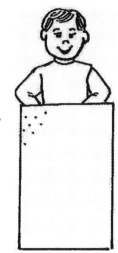

KNOWLEDGE

Locate the words to a given speech from history. Consider such speeches as Abraham Lincoln's "Gettysburg Address," FDR's "The only thing we have to fear is fear itself" speech, or John F. Kennedy's "Ask not what your country can do for you, but what you can do for your country," speech. Read through them carefully.

COMPREHENSION

In your own words, explain what makes these famous speeches so great. Consider their purpose, vocabulary, use of figurative language, historical time period/event, speaker's passion, intended audience, and place where they were given.

APPLICATION

Practice reading one or more of these speeches aloud, perhaps as a choral reading with some of your peers. Assign parts to individuals and small groups. Consider when it is best to pause, when to use voice inflections, when to speak softly/loudly, when to enunciate selected words, and anything else to add drama.

ANALYSIS

It has been said that these famous speeches, along with many others, are speeches of intense passion. Point out the most poignant or passionate passages in one or more of the speeches.

SYNTHESIS

Think of a subject or controversial issue that you feel passionate about such as war, peace, abortion, stem cell research, death penalty, drug testing, same sex marriages, divorce, right to vote, etc. Try writing an emotional speech that conveys your strong feelings on whether something is right or wrong and what should be done about it!

EVALUATION

Research suggests that one of the things people fear most is having to give a speech in front of an audience. Why do you think this is true?

Ignite Student Intellect and Imagination in Language Arts, by Sandra L. Schurr and Kathy L. LaMorte. Published by National Middle School Association, 2007.

PART II
Imaginative Assessment Options and Real-Life Applications

The term "authentic assessment" has claimed a place in the lexicon of education as the inadequacy of paper and pencil tests was more widely recognized and other means of measuring student progress were developed. These measures call for a more genuine or lifelike demonstration of mastery, hence are labeled "authentic."

Such demonstrations can be accomplished through the use of a portfolio, through the design and construction of a product, or through a student-generated demonstration or performance. The levels, behaviors, and verbs of Bloom's Taxonomy provide the teacher and student alike with a variety of experiences that include everything from work samples, self-assessment ratings, and observations to simulated performances, interviews, and creative presentations. In essence, the taxonomy provides a built-in process for measuring what students know and can do through a series of products, performances, and varied artifacts for a portfolio requirement.

Teachers and students can use these two final sections in the book as extensions of the Bloom Sheets in multiple ways and for multiple purposes. The ideas in the first section can provide alternative or additional tasks to demonstrate proficiencies in both content and skills. Likewise, the suggested research topics can serve as further enrichment for teachers to assign or students to select as part of the course outcomes on an individual or class need basis.

The Real-Life Applications pages provide important tools for the teacher to use in showing students the relevance of language arts. Students all too seldom see this transfer of knowledge on their own, so teachers should help build bridges between theory and life experiences wherever possible. What makes these last two sections even more valuable, of course, is the fact that they, too, are related to standards for the English language arts.

110

Assess Learning by
Putting Language Arts Artifacts Into Your Portfolio

- A language arts quiz, test, or exam.

- A language arts homework assignment.

- A language arts textbook set of questions and their corresponding answers.

- A language arts glossary of terms you have developed.

- A language arts annotated bibliography of print materials or Internet Web sites.

- A personal reaction to a newspaper feature story, news story, editorial, or editorial cartoon.

- A language arts-related rubric or self-checklist.

- A list of language arts-related resources in your community.

- A cooperative learning language arts activity that you played a major role in completing.

- A language arts learning station activity or assignment that you enjoyed.

- A copy of an original speech that was written and delivered in class.

- A piece of original writing in the form of a short story, play, poem, biography, or autobiography.

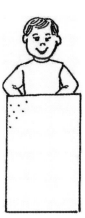

Ignite Student Intellect and Imagination in Language Arts, by Sandra L. Schurr and Kathy L. LaMorte. Published by National Middle School Association, 2007.

Assess Learning by
Creating Language Arts Products

- A report or position paper on a topic of choice.

- A set of field notes from a trip or excursion.

- A lesson plan to teach others about something in language arts.

- A how-to booklet with directions on how to perform a task or complete a project.

- A chart or graph representing a data collection/analysis task or assignment.

- A set of language arts-related fact cards or flash cards.

- A mobile, poster, collage, or other unusual report format based on an assigned topic.

- A role play or case study on a language arts-oriented setting, problem area, or dilemma.

- A series of reflective journal or learning log entries related to assigned readings.

- A crossword or word finder puzzle related to a topic of choice.

- An annotated bibliography of print resources or Internet Web sites related to language arts.

- An audio or video tape of an interview, panel discussion, or debate.

Ignite Student Intellect and Imagination in Language Arts, by Sandra L. Schurr and Kathy L. LaMorte. Published by National Middle School Association, 2007.

Assess Learning by
Delivering Language Arts Performances

- A language arts display, exhibit, or demonstration including media techniques.

- A commercial promoting language arts throughout the school community.

- A chalk talk or mini-speech on a topic of choice.

- A tape recording of a lesson you taught to classmates.

- Participation in a debate on a relevant topic.

- An interview with a local author or poet.

- Plans and implementation for a language arts-related contest

- A book report or review.

- A promotion and outline for a language arts fair project.

- A Reader's Theater presentation of a play, essay, biography, or story.

- Participation in a choral reading activity.

- Demonstration of your ability to use a variety of technological and informational resources available to you.

Ignite Student Intellect and Imagination in Language Arts, by Sandra L. Schurr and Kathy L. LaMorte. Published by National Middle School Association, 2007.

Assess Learning by
Conducting Research on Language Arts Topics

- Literature from other cultures.

- Elements of contemporary versus classic literary works.

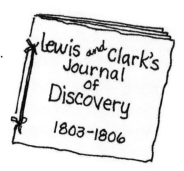

- History of the alphabet.

- Other alphabets.

- Literature from different periods in history.

- Different writing process elements.

- Varied language structures and conventions.

- Biographies of famous authors and poets.

- Research and position papers on current societal trends and issues.

- Technological and informational resources (libraries, databases, computer networks, videos) available for today's students.

- Diversity in language use, patterns, and dialects across varied ethnic groups and geographic regions.

- Use of spoken, written, and visual language effective with ESOL students.

- Body language and its implications for communication.

Ignite Student Intellect and Imagination in Language Arts, by Sandra L. Schurr and Kathy L. LaMorte. Published by National Middle School Association, 2007.

Real-Life Application
The Picture Book as a Literary Source

DIRECTIONS: Browse through the children's section of your community library and notice how books come in all shapes, sizes, and formats. Also note how children's books can be read and interpreted on two levels—a literal level and an interpretive level. Like comic strips, readers of all ages can enjoy the story line at its most basic level or at a deeper level of understanding. In this activity, you will create a picture book for students in elementary grades. It can reflect any historical period, any genre, and any fiction or nonfiction topic.

1. Read a variety of picture books at the elementary level to note how they use both script and illustrations to get their story line across. Also notice that these books are very sophisticated, informative, and versatile covering many different subject areas, genres, and historical periods.

2. Decide on the type of picture book you would like to write. Will it be fiction or nonfiction? Will it be a book of poems, short stories, or one longer story?

3. Decide on a specific topic or theme for your book. Then begin the writing process to complete the book's content.

4. Once the story line is finished, decide how you will illustrate the book. Will you draw the pictures yourself, recruit someone in class to illustrate it, or download pictures from the Internet? How will you break down the book's copy to make the best use of the illustrations?

5. Now it is time to make a decision on how to format the book. Here are some ideas to consider:
   (1) A pop-up book
   (2) A flap book
   (3) An accordion book
   (4) An ABC book
   (5) A roll movie book
   (6) A spiral bound book
   (7) A book on tape.

Use the Internet to locate several Web sites for creative bookmaking to help you make this decision and to give you directions on additional formatting options.

Ignite Student Intellect and Imagination in Language Arts, by Sandra L. Schurr and Kathy L. LaMorte. Published by National Middle School Association, 2007.

Starting a Teenage Book Club

DIRECTIONS: Book clubs have become popular literary and social events over the past several years thanks in part to their early sponsorship and promotion by the "Oprah Winfrey Show" and the "Today Show." Now young adult or teenage book clubs are springing up in middle schools meeting both in and out of the traditional school day. Follow these steps for organizing a book club structure for your friends or school.

1. Go online and visit a popular book club Web site such as Amazon.com to find out more details about starting or belonging to a book club at either the local or national levels. What advice and guidelines do they provide their viewers? How are books selected? What purposes do book guides serve in the process? Who conducts the book club discussions? What is the ideal size for a club? When and how often do these groups meet?

2. Work with a small group of interested peers to generate a list of criteria for a book that makes a "good read." Then, write down a list of books that your group has read and that meets the designated criteria.

3. If time permits, take turns developing a discussion guide of questions for each of the books listed above. Encourage readers to use the questions as an outline for analyzing the character, setting, plot, and literary qualities of these books.

4. Talk with your language arts teacher or media specialist to investigate the possibility of starting a young adult book club at your school. Sometimes these are part of the school day in lieu of study hall or sometimes they are co-curricular and held after school hours. In some cases, they can even be held in the evenings at the homes of group members on a rotating basis using an interested parent as the sponsor and/or facilitator.

5. Note that most successful book clubs limit their membership, have strict guidelines for attendance and participation, and vary book selections to cover a wide range of genres, authors, and historical periods.

Ignite Student Intellect and Imagination in Language Arts, by Sandra L. Schurr and Kathy L. LaMorte. Published by National Middle School Association, 2007.

Real-Life Application
Organizing a Letter-Writing Campaign

DIRECTIONS: One can accomplish a great deal in this world by learning to write effective letters. After all, "the pen is mightier than the sword." This activity will enable you and a group of peers to organize a letter-writing campaign to accomplish a specific purpose in making a difference in your community.

1. Spend some time reading the Letters to the Editor in your local newspaper. Notice what issues, causes, subjects, and controversial topics are covered in this letter-writing effort by members of your local community.

2. Note that there are several different types of letters to consider when trying to persuade people to action.
    - There are informational letters that gather information or that pass on information to someone else.
    - There are support letters that let people know you agree with them.
    - There are persuasive letters that try to influence someone to do something.
    - There are opposition letters that let people know when you don't agree with them.
    - There are problem/solution letters that identify a problem issue and then propose one or more solutions to the problem.
    - There are request letters that ask for someone's help or encouragement.

3. Determine what type of letter-writing campaign your group of peers would like to undertake. Establish a goal or mission to accomplish as well as a plan of action. Outline roles and responsibilities for each group member as well as a time line for completing the task.

4. Let the "letters begin" and document the results of your letter-writing efforts.

Ignite Student Intellect and Imagination in Language Arts, by Sandra L. Schurr and Kathy L. LaMorte. Published by National Middle School Association, 2007.

Real-Life Application
Writing to Promote Thyself

DIRECTIONS: Life skills are becoming an integral part of the schooling process to better prepare kids for today's workplace. More and more students are choosing to get jobs as soon as they reach the legal age of 16. This set of activities is designed to help students better understand the purpose and nature of effective resumé writing.

1. Use the Internet to locate information from several Web sites that address one or more of the following job-related topics:
   - What percentage of eligible teens are active in today's workplace, both during the school year and in the summer?
   - What types of jobs are most plentiful for these kids? Most popular? Most desirable? Most profitable? Most skill-based? Clock the most hours?
   - What qualifications are most important for a teenage worker to have?
   - What do most kids do with their earnings—save or spend?
   - What kinds of training and safety standards are most typical of these jobs?
   - What encompasses a typical job interview process?

2. Next, collect examples of several job applications from popular teenage employers in your community. What do they all seem to have in common?

3. Practice role playing the job interview process with a peer. Take turns being the employer and the prospective employee.

4. Seek examples of quality resumés from self-help books in a library or bookstore as well as online. Note what makes one example better than another. Compose a resumé for yourself at this time. Include all relevant information such as:
   - Your name, address, telephone number, and personal Web site
   - Your academic grades and accomplishments or special skills/talents/interests in school
   - Your extracurricular activities and interests outside of school
   - Your personal goals and work expectations
   - A list of personal references who would vouch for your character.

5. For fun, use the computer to format and print out a personal business card to distribute when you go job seeking or when you are out and active in the community. Create a logo or slogan to represent your character and personality.

Ignite Student Intellect and Imagination in Language Arts,
by Sandra L. Schurr and Kathy L. LaMorte. Published by
National Middle School Association, 2007.

Putting On a Drama Festival

DIRECTIONS: Show off the talents in your school by planning and implementing a unique and intimate Drama Festival. Plan this event for a small group setting and in a cozy atmosphere if possible. The program could be presented several times to different groups of peers, parents, or interested parties in the community.

1. Select a time, date, and place for this event. Promote the program through printed invitations, publicly displayed posters, and/or through individual e-mail messages.

2. Recruit individuals to participate in one of these four types of performances:
   Reader's Theatre
   Skits or plays
   Poetry readings
   Monologues

3. Individual participants may write their own original scripts for these performances or they can search for quality poems, plays, stories, and monologues that would be appropriate to study and use as material for this drama festival.

4. Provide time for the participants to memorize and/or rehearse their material for the festival. Remember that practice makes perfect.

5. Think of ways to enhance the delivery of each performance. Consider setting, lighting, background scenery, soft music, props, and anything else to set the tone of intimacy and drama.

6. Determine how you will "assess the success" of your drama festival. Consider attendance numbers, reactions/comments from audience, thoughts/feelings from participants, and anything else that matters.

Ignite Student Intellect and Imagination in Language Arts, by Sandra L. Schurr and Kathy L. LaMorte. Published by National Middle School Association, 2007.

Real-Life Application
Producing a Photo Shoot

DIRECTIONS: Digital photography has revolutionized the picture taking world. This new technology is to be the foundation for a research-based project. You will use a digital camera to conduct a photo shoot on a topic of your own choosing. So let's see the lights—camera—action!

1. Get to know your local community better by thinking of a topic or subject area that is of special interest to you or that you would like to know more about. Consider focusing on a local park or beach, museum, tourist attraction, historical site, unusual architecture, or unique things to see and do in your community.

2. Once your subject has been chosen, visit local Web sites on the Internet to collect and record important data about the subject.

3. Next, think of how you might best portray your subject through pictures as well as words. Locate a digital camera and learn how its technology works. To get you started, note these things that are unique to a digital camera as compared with a camera dependent upon film:
   • Digital camera records photos on memory cards, not film.
   • Computer reads the memory card stored in a card reader and plugged into the computer.
   • You can look at the photos, e-mail the photos, and/or store photos on the hard drive.
   • Memory cards allow you to print photos or put on CD.

4. Now, outline the information you have electronically collected on your topic and use to sequence your important ideas/facts. Then, think of a photograph you might take to portray each major idea in your outline.

5. Complete your photo shoot by taking the appropriate pictures and editing and cropping them before printing them out. You might even want to put them on a CD.

6. Prepare your final script of information from the outline so that you have a quality paragraph or two for each designated photograph.

7. You may place each photograph with its corresponding paragraphs in an album or you can prepare a script to be read when you show your photographs on a CD one by one.

Ignite Student Intellect and Imagination in Language Arts, by Sandra L. Schurr and Kathy L. LaMorte. Published by National Middle School Association, 2007.

Be a Part of an ESOL/Diversity Assistance Team

DIRECTIONS: Today's schools are exciting melting pots that serve many diverse ethnic groups. A significant number of these students require ESOL assisted instruction from the teacher as they learn the required subject area and content for math, science, language arts, and social studies. The following list of activities can help you and the teacher tutor or assist these special needs kids in the classroom while you are learning new skills and concepts as well.

1. It is important that both students and teachers understand that there are key cultural characteristics which may affect parent/student/teacher communication efforts on a daily basis in the school. Some of these things to keep in mind are:

    AFRICAN AMERICANS
    . . . Have a deep sense of cultural history
    . . . Are physically expressive with gestures and body language
    . . . Display emotional intensity and expression during conversation.

    HISPANICS
    . . . Seek closeness during conversation
    . . . Have flexible sense of time
    . . . Feel eye contact is important during conversation.

    ASIAN AMERICANS
    . . . Are reserved with a respect for silence and control
    . . . Drop eyes to show respect
    . . . May take offense at use of first names only and shaking hands with a person of the opposite sex.

    ANGLO AMERICANS
    . . . Value direct and polite conversation
    . . . Are competitive and individualistic
    . . . Are oriented toward privacy with disdain for public reprimand.

2. Try working with your teacher to assist him or her as a co-worker in completing one or more of these instructional strategies that help ESOL students better understand the teacher's lectures and directed teaching lessons:
    *   Provide a handout or outline summarizing major points of a lecture but instead of providing complete text, leave some portions blank for students to fill in. Examples: Leave either term or definition blank; leave several items in list blank; leave key words in statement blank; leave first or last names of key people blank.

- Ask short, but relevant questions throughout lecture that require students to respond by voting. Examples: How many of you understand that definition? Are we ready to move on? How many still have questions about this topic?
- Write key words, ideas, numbers, symbols on chalkboard during lecture using colored chalk to highlight major ideas.
- Appoint selected students as chalkboard note-takers during a lecture to help students who have difficulty hearing, seeing, or writing and listening at the same time.
- Use choral work after making an important point in a lecture. Hold up king-size prompts or flashcards and ask students to read aloud in unison what's written on the card several times. At end of the lecture, review concepts by holding up all cards used during the time period.
- Use the Whip Around/Pass Option with a small number of students by asking them to respond to an idea in the lecture through completion of a starter statement such as: "I agree with that idea because" or "I also read that" or "I thought that."
- Use Response Cards during a questioning session that follows a lecture.

3. Pair up with an ESOL or other student for one of these 10 suggested assignments that benefit learning partners:
   - Discuss a short written document together.
   - Interview each other concerning partner's reactions to an assigned reading, lecture, video viewing, or any other assigned task.
   - Critique or edit each other's written work.
   - Question your partner about an assigned reading.
   - Recap a lesson or class session together.
   - Develop questions together to ask the teacher.
   - Analyze a case problem, exercise, or experiment together.
   - Test each other.
   - Respond to a question posed by the teacher.
   - Compare notes taken in class.

Ignite Student Intellect and Imagination in Language Arts, by Sandra L. Schurr and Kathy L. LaMorte. Published by National Middle School Association, 2007.

Real-Life Application
Starting a Junior Toastmaster Club

DIRECTIONS: The goal of the National Toastmaster's Club is to improve member communi–cation (speaking and listening) and leadership (thinking) skills. Members of the club learn by speaking to groups and working with others in a supportive environment. A typical Toastmasters Club is made up of 20 to 30 people who meet once a week or twice a month for an hour or so. Each meeting begins with a short business session which helps members learn basic meeting procedures. Next, each member gives short impromptu speeches on assigned topics. The final activity provides time for three or more members to present prepared speeches on such topics as speech principles and organization, voice, language, gestures, and persuasion techniques. It is important to note that every prepared speaker is assigned an evaluator who points out speech strengths and offers suggestions for improvement. Follow these guidelines to start your own school-related junior Toastmaster's Club.

1. Find a teacher sponsor to facilitate your group. Decide on a time, place, and dates for meetings to take place. Recruit students who are interested in improving their speech writing and delivery skills as well as their leadership qualities.

2. Begin building a quality library of tools related to "speaking" and "speech making." Consider manuals, books, audio/visual cassettes, CD's and an annotated bibliography of relevant Internet Web sites.

3. Set up a procedure for each meeting that includes
   - Rotating leadership for conducting business session of meeting.
   - Preparing and randomly assigning topics for impromptu speeches for all members present.
   - Requiring two or three members to prepare and deliver a speech at each meeting on appropriate speech-related topics.
   - Providing time for evaluation of prepared speeches.

Ignite Student Intellect and Imagination in Language Arts, by Sandra L. Schurr and Kathy L. LaMorte. Published by National Middle School Association, 2007.

Bringing Humor to the Language Arts Classroom

DIRECTIONS: There are many benefits to promoting humor in the classroom. They are: (1) It fosters friendship, bonding, and building of student/teacher relationships both emotionally and psychologically. (2) It provokes thought by forcing students to make connections they might otherwise miss; (3) It reduces anxiety because one can't laugh and be depressed at the same time. (4) It enhances creative thinking by giving us new ways to look at familiar objects and events; (5) It aids in the retention of material by making learning more enjoyable; (6) It relaxes and engages students by releasing them from fear and apprehension; and (7) It relieves monotony and boredom by helping students stay tuned in. Complete the activities suggested below to help you bring more humor and laughter into your language arts classroom!

1. Determine your humor quotient by answering YES or NO to each of these statements:
   - My friends describe me as a funny person at home and at school.
   - My friends and family think my sense of humor is one of my strongest assets.
   - My humor is not negative and degrading.
   - I can laugh at myself and my mistakes.
   - I sometimes laugh alone when I see or hear something funny.
   - I laugh and enjoy jokes and stories.
   - I often seek out funny shows, cartoons, and books.
   - I always look at the funny side of life.
   - I send humorous cards, notes, and e-mails to my friends.
   - My sense of humor makes it hard for people to get mad at me.
   - I sometimes act silly when people least expect it.
   - I use humor to help me remember or recall information.

2. There are 15 stages of humor that have been identified by Kuhn. They are in order from least funny to most funny: *smirk, smile, grin, snicker, giggle, chuckle, chortle, laugh, cackle, guffaw, howl, shriek, roar, convulse,* and *die laughing* (Kuhn, C. C. [1944]. The stages of laughter. *Journal of Nursing Jocularity, 4* [2], 34-35). Tell a funny joke to an audience or group of peers. Rate their reaction to your joke on these 15 stages of laughter.

3. Use each of the "humorous springboards" which follow to create an original adaptation of your own. These "good for a laugh" examples provide you with many educational skill and concept applications which include: (1) Sentence construction and use of dialogue; (2) Knowledge of parts of speech; (3) Introduction of alliteration, synonyms, homophones, rhyming patterns, metaphors, similes, and exaggeration; (4) Creation of playful imagery; (5) Vocabulary development; (6) Dictionary and study skills; (7) Increased language usage; and (8) Creative thinking.

HUMOR SPRINGBOARD 1: Daffy Definitions
Examples: Archeology: digging up the past
              Justice: what we get when the decision is in our favor

HUMOR SPRINGBOARD 2: Puns

Examples: The best place to buy electrical appliances is at an outlet store.

A single shoe store owner is called a sole proprietor.

HUMOR SPRINGBOARD 3: Alliteration on Famous People

Examples: Franklin's friends in France found famous Franklin fierce but friendly.

Presidents Pierce and Polk probably ate popped popcorn.

HUMOR SPRINGBOARD 4: Riddles

Examples: How do famous cars sign their names? With autographs!

What kind of deer always carries an umbrella? A reindeer!

HUMOR SPRINGBOARD 5: Stink Pinks

Examples: What do you call a lawful bird? A legal eagle

What do you call a dieting 16th President? A shrinkin' Lincoln!

HUMOR SPRINGBOARD 6: Paraphrased Proverbs

Examples: The early bird catches the worm.

The prompt feathered friend overtakes the creepy crawler.

HUMOR SPRINGBOARD 8: Picture and Draw These Idioms

Examples: He is all thumbs.

She has butterflies in her stomach.

He laughed his head off.

What's the matter? Does the cat have your tongue?

HUMOR SPRINGBOARD 9: Tom Swifties

Examples: "Get to the point," he said sharply.

"These hot dogs are great," she said frankly.

"I am afraid of canines," he said doggedly.

"Don't touch that burner," she said hotly.

HUMOR SPRINGBOARD 10: Stately Clues

Examples: Think of a real city in Wyoming that is not Pepsiville. Answer: Cokeville

Think of a city in Wyoming that is a friendly ghost. Answer: Casper

Think of a city in Wyoming that could be Razor City. Answer: Gillette

4. Try building a personal file of humorous items by collecting funny examples of these types of items: Jokes, Riddles, Quotations, Comic Strips, Editorial Cartoons, Limericks, Bumper Stickers, Tongue Twisters, Advertisements, Puns, Exaggerations, Maxims or Proverbs, Murphy's Laws, Daffynitions, Clichés, Idioms, and Brain Teasers.

It's raining cats and dogs!

Ignite Student Intellect and Imagination in Language Arts, by Sandra L. Schurr and Kathy L. LaMorte. Published by National Middle School Association, 2007.

PART III
More Instructional Tools and Techniques

With middle level students the need to differentiate instruction is clear. Their diversity demands it. In language arts classrooms that need is almost compelling, for language arts instruction has often suffered from the overuse of a staid pattern featuring grammar worksheets and reading aloud.

In this last section, a variety of tools and techniques are presented that will take grammar out of the workbook and literature out of the anthology and engage students in different ways, all of which will lead them toward acquiring that meaningful understanding of the English language and its uses in communication that is universally sought. Students will come to appreciate language more as they carry out some of the imaginative tasks suggested. These activities will more adequately prepare them for the standardized tests that have become the prime way to demonstrate achievement.

The first two pages offer 10 varied ways to use Bloom's Taxonomy in the language arts classroom. One of the ways calls for using the taxonomy as a basis for constructing tests that the student will find appealing. In addition, there is a separate list of 10 creative ways to assess student progress, a list of ways to do research reports on language arts topics, and one on ways to improve questioning, plus other lists. Take time to explore these imaginative possibilities for improving learning.

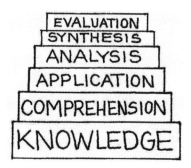

1. Ten Smart Ways to Use Bloom's Taxonomy in the Language Arts Classroom

1. Set up permanent **Learning Stations or Centers** around the classroom and label each station according to the levels of Bloom's Taxonomy: Knowledge Station, Comprehension Station, Application Station, Analysis Station, Synthesis Station, and Evaluation Station. Be sure to put the major verbs or behaviors associated with each level on the Learning Station Signs as cue words and reminders for students on what type of thinking they are using while working at each station task or set of tasks.

2. Create a **Mini-Interdisciplinary Unit** based on Bloom's Taxonomy. Decide on a universal theme and then incorporate as many of the subject areas as you can for each level of Bloom. Organize your interdisciplinary tasks so that all knowledge level tasks are together, all comprehension level tasks are together.

3. Use Bloom's Taxonomy as the organizing structure for doing **Internet Research.** Encourage students to develop a research outline of six levels—one for each level of Bloom. Students choose a topic to study and then generate a series of questions and answers on that topic for each level of the taxonomy.

4. Use **Commercial Posters** that focus on any language arts-related topic as a springboard for developing a set of questions or tasks about the content of the poster—one for each level of Bloom's Taxonomy. Commercial posters may be obtained from travel agents, book stores, teacher stores, magazine subscriptions, or by having kids bring in their favorite posters from home.

5. Construct a set of **Investigation Cards** on an academic theme or topic of choice. There are 18 task cards in a set—three for each level of Bloom's Taxonomy. Each card has a different question or task related to the overall topic/theme for students to complete. Teachers can assign specific cards to students or students can select their own cards to do. Teachers might also suggest that students complete at least one card at each level of the taxonomy or they can assign specific levels for kids to focus on. Finally, teachers can assign a set of cards to cooperative learning groups, with each group having the same set of cards, or each group working on a different set of cards.

6. Design a **Six Question Quiz or Test** around Bloom's Taxonomy. Include just one question for each level of the taxonomy and assign points to each one so that the total test is worth 100 points. Example: Knowledge Question = 5 points; Comprehension Question =10 points; Application Question = 15 points; Analysis Question = 20 points; Synthesis Question = 25 points; and Evaluation Question = 25 points. Imagine how kids will like knowing that their test will only have six questions to answer!

7. Design **Multiple Question Quiz or Test** around Bloom's Taxonomy. Develop a set of at least five questions for each level of the taxonomy and assign points to each one as you did in the previous quiz/test format. Instruct students to select any combination of questions to answer as long as their total point value equals 100 points. Imagine how kids will like choosing their own test questions to answer!

8. Create **Bloom Sheets** of your own. Use computer graphics, stencils, geometric shapes, or coloring book outlines to illustrate your questions and tasks.

9. Develop a **Discussion Organizer** for any classroom discussion group around Bloom's Taxonomy to ensure that you are including both low and higher order thinking skills in the dialogue. Prepare several questions at each level of the taxonomy in advance of the large group discussion to keep things interactive and interesting.

10. Assign both fiction and nonfiction **Book Reports** to students requiring them to answer at least one question about their book for each level of Bloom's Taxonomy. Prepare and distribute these questions prior to the student's reading of the book which can then serve as both a reading guide and as a report format.

Ignite Student Intellect and Imagination in Language Arts, by Sandra L. Schurr and Kathy L. LaMorte. Published by National Middle School Association, 2007.

2. Ten Unique Ways to Research and Write a Report for Language Arts

1. Write an ABC research-based report for any language arts topic that consists of a comprehensive paragraph describing or discussing a key concept or topic for each letter of the alphabet.

2. Write a report that uses a set of fact cards for its presentation of language arts-related information. On one side of each card, compose a paragraph of description and on the other side draw or use computer clip art to feature a graphic correlated to the card's content.

3. Design a report in the form of a wall calendar for language arts geeks. Locate a language arts-oriented picture, diagram, chart, graph, or topic from the Internet for each month of the year and then create a summary paragraph explaining what it is or does.

4. Write a report that uses bookmarks as a format for recording information about important language arts concepts. Choose a different language arts concept for each bookmark and write down factual statements, definitions, examples, or applications of each concept. Use these to "mark your place" in a language arts-oriented text or reference book!

5. Compile a time line report that centers around the important dates in the life of an author or poet. Choose a person to research and then use a series of key dates for his/her accomplishments. Be sure that each date includes a detailed description or explanation to go with it.

6. Put together a graphic organizer report that summarizes information that you have gleaned from the Internet or varied print/nonprint resources on a language arts-oriented topic of special interest to you. Some graphic organizers to use as part of this process are KWL Chart, Word Web, Venn Diagram, Ladder, Comparison and Contrast Organizer, Mind Map, Chain or Cycle Graph, Tree, Fishbone, or Concept Map.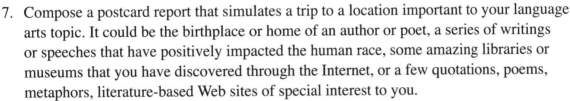

7. Compose a postcard report that simulates a trip to a location important to your language arts topic. It could be the birthplace or home of an author or poet, a series of writings or speeches that have positively impacted the human race, some amazing libraries or museums that you have discovered through the Internet, or a few quotations, poems, metaphors, literature-based Web sites of special interest to you.

    Design the set of postcards so that they feature a drawing on one side of the card with a descriptive message/address on the other side of the card. Try creating a special postage stamp to commemorate the person, place, or concept as well.

8. Create a placemat report that focuses on a language arts topic you need to study in some detail. Parts of speech, grammar and spelling rules, alternative sentence structures, famous authors, literary techniques, or varied genres are all good topics for this type of format.

9. Assume the role of a newspaper editor and use this format as a way of recording information for a language arts-related report. Write a feature story, news story, editorial, classified advertisement, comic strip, or book review that mimics a real newspaper layout.

   Editing Checklist
   ✓ Content & Organization
   ✓ Style
   ✓ Format

10. Create a pictorial report on an interesting fiction or non-fiction book that uses a series of photos, diagrams, illustrations, pictures, or charts to tell its story. Put each graphic on a page with a summary paragraph to accompany it. Combine your pages into a simple booklet with a Table of Contents and a Glossary if needed.

Ignite Student Intellect and Imagination in Language Arts, by Sandra L. Schurr and Kathy L. LaMorte. Published by National Middle School Association, 2007.

3. Ten Alternate Ways to Measure
What Students Know and Can Do in Language Arts

1. **Fact/Statistic/Example Exercise.** Instruct students to use textbooks or reference books to locate and record important facts, numbers/statistics, and examples relevant to an important topic. Remind them that a *fact* is something that can be verified or validated as being real; that a *statistic* is numerical information or information that is explained using numbers; and that an *example* is specific information or data that can be used to represent a group as a whole.

2. **Jeopardy Review.** Divide an 8½ x 11-inch piece of paper into 9 or 12 equal sections or squares. Instruct students to divide each large square into two parts and label the top part *answer* and the bottom part *question*. Give each student a key language arts concept to place below the *answer* space and have them generate a quality question for that particular answer in the *answer* space Jeopardy style.

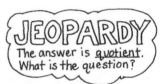

3. **Tic-Tac-Toe Game.** Divide the class into groups of three. Provide each group with a set of Tic Tac Toe diagrams that have three different language arts-oriented questions in each of the nine cells. Assign two of the students to be players in the game of Tic-Tac-Toe (one being X and one being O) and the third player to be the referee with the textbook or answer key. Students play the game of Tic-Tac-Toe taking turns and selecting a question to answer. The referee accepts or rejects each player's response and placement of an X or O. Students take turns being the referee and the players.

4. **Internet Research.** Ask students to research a topic on the Internet. Instruct them to create three questions or tasks for each level of Bloom's Taxonomy to guide their recording of information. Students generate these questions/tasks as they read and learn about their topics. Students should then complete their answers or tasks as part of the learning process. Have them discuss their experiences by focusing on such ideas as: Which questions had right or wrong answers? Which questions were easiest? Most challenging? Most difficult? Most thought-provoking?

5. **Response Cards.** Provide each student with two different sets of feedback cards. The first set should have *true* on one side of the card and *false* on the other side of the card. The second set should have an A on one side of the card and B on the other side of the card with a C on still another card and D on the reverse side of that card. The teacher then proceeds to make a series of *true/false* statements about a language arts-oriented topic/concept/ skill followed by a series of *multiple choice* questions, and the entire class holds up the appropriate card for each type of question presented.

6. **Students as Teachers.** Hand out a blank index card to each student. Ask each student to write down any questions he or she may have about a language arts skill, topic, or concept being studied. Collect the cards, shuffle them, and distribute one to each student. Ask students to read silently the question or topic of their card and think of a response. Invite volunteers who are willing to read out loud the card they obtained and give a response. After a response is given, ask the others in the class to add more information/explanation to what the volunteer has contributed.

7. **Hollywood Squares.** Simulate the tic-tac-toe game show format used on Hollywood Squares. Ask three volunteers to sit on the floor in front of the chairs, three to sit in the chairs, and three to stand behind the chairs. Give each of the nine "celebrities" a card with an X printed on one side and an O on the other side to hold against his or her body as language arts-related questions are answered successfully. Ask for two volunteers to serve as contestants. The contestants pick members of the "celebrity" squares to answer the game's questions. Ask the contestants to take turns. The contestants respond with *agree* or *disagree* to the panel's response as they try to form a winning tic-tac-toe.

8. **Concept Relationship Tests.** The language arts teacher identifies nine important language arts terms or concepts that have been taught on a topic and requires each student to write each of these down on a file card—one idea per card. The student then shuffles the cards and randomly lays them out with three cards across and three cards down so that there are three cards vertically, horizontally, and diagonally. The student is then told to write a single statement for each group of three cards across, down, and diagonally showing the relationship or interfacing of the appropriate cards to one another.

9. **Panel Discussion.** Divide students into groups of five. Designate four as panel members and one as the moderator for each group. Once a language arts topic has been selected, each panel member must choose a specific part of the topic on which to become an expert. Panel members then prepare two to four questions that they are prepared to answer and turn these in to the moderator. The moderator's job is to organize the panel's questions and then ask the questions during the presentation with enough background information to keep the discussion going smoothly. At the end of the discussion, the moderator summarizes the key points for the audience that have been shared by the group.

10. **Fact-Finding Sharing Tests.** The teacher should prepare a classroom set of information cards on a language arts topic of his or her choosing. Each card must be different and must contain a short paragraph describing an important term, concept, or issue related to the topic. For example, copying passages from a textbook works well for this activity. Duplicate cards may be developed for concepts or passages that are more difficult. Each student in the class is given a different information card and told to read it silently. Once the student understands the information given, he or she is ready to teach it to others. The teacher then has students stand and circulate among one another, pausing just long enough to informally share and retell the information to another student or small group of students. Each person tries to both "teach" and "learn" new information using this procedure. After

approximately 30 minutes of such exchanges, the teacher asks the students to return to their desks and respond to these five starter statements:

(1)   Important information on my card that I had to teach others about the topic was…
(2)   Three new things I learned from other students on the topic were…
(3)   Something about the topic that I would like to know more about is…
(4)   One thing about the topic that I already knew was…
(5)   To me, the most interesting thing about this topic is…

The teacher should then collect the student response sheets and determine what has been learned through this activity and what needs to be taught/reviewed again through a different method. It is important that students develop good listening and speaking skills if this strategy is to be effective. Students may not read from their information cards directly, but can use them as reference cards in retelling the information in their own words.

Ignite Student Intellect and Imagination in Language Arts, by Sandra L. Schurr and Kathy L. LaMorte. Published by National Middle School Association, 2007.

4. Sixteen Types of Questions to Challenge a Student's Ability to Think Like a Reading and Writing Expert

NOTE: Use these 16 types of questions and their corresponding examples as an outline for conducting research on any language arts-related topic or subject area.

Analyzing: What qualities/characteristics do you think are most important for an author or poet to possess?

Applying: What practical applications can you think of for demonstrating one's knowledge and understanding of the parts of speech?

Comparing/Contrasting: What similarities and differences exist between and among the genres of literature?

Defining: How would you define and describe these types of folklore: legends, tall tales, and fairy tales?

Evaluating/Assessing: What are the advantages and disadvantages of developing rubrics to critique your writing assignments?

Explaining: How could you explain the different writing process elements to a younger student?

Generalizing: What are the major causes of spelling and punctuation errors in your writing assignments?

Investigating: How could you find out more about a given author or poet?

Patterning: What types of rhyming patterns are found in various forms of poetry?

Predicting/Hypothesizing: How can one predict the ending to a play, story, or novel?

Problem Solving: What are some possible ways to organize the structure and body of a speech?

Questioning: What questions do you have about the relationship between parts of speech and the diagramming of sentences?

Reducing/Simplifying: How can you reduce or simplify the use of a variety of technological and informational resources to gather and synthesize information in order to communicate knowledge?

Reflecting: How has your thinking changed about the way to conduct research on issues and problems for a research paper?

Relating: What is the relationship between effective note cards and the writing of a quality position paper?

Sequencing: What steps are involved in building a complex sentence?

Ignite Student Intellect and Imagination in Language Arts, by Sandra L. Schurr and Kathy L. LaMorte. Published by National Middle School Association, 2007.

NOUN
abstract
common
collective
proper
compound
concrete

5. The Art and Science of Asking Effective Questions in the Language Arts Classroom

1. Always plan classroom questions in advance of the discussion to avoid leaving something important out of the content being reviewed.
2. Vary the cognitive level of your questions according to Bloom's Taxonomy.
3. Avoid asking too many questions too fast to discourage shallow answers.
4. Do not repeat questions routinely as this trains students not to listen attentively the first or even the second time around.
5. Provide think time of 5 to 15 seconds after most questions to improve quality of student responses.
6. State questions before calling on students for the answer so as to hold everyone's attention, not just the attention of the student addressed.
7. Do not repeat student answers as this will diminish student incentive to listen to one another the first time.
8. Do not add information to improve student answers initially, but instead give them clues so they can embellish or build on their own ideas.
9. Call on both volunteers and non-volunteers to keep everyone involved.
10. Make questions opportunities for students to take chances and learn as opposed to punishments for inattentive students who don't listen or value taking risks.
11. Encourage students to direct answers to one another and not just to the teacher. This promotes the teacher's role as "guide on the side" rather than "sage on the stage."
12. Encourage students to question each other whenever possible to do so for purposes of promoting open dialogue between and among students.
13. Honor student questions as part of the ongoing discussion rather than rely only on those topics of interest to the teacher.

Ignite Student Intellect and Imagination in Language Arts, by Sandra L. Schurr and Kathy L. LaMorte. Published by National Middle School Association, 2007.

6. Twelve Creative Writing Prompts Related to the Language Arts Standards

Standard One

Madeleine L'Engle, author of *A Wrinkle in Time,* once wrote: "Reading is a creative activity. You have to visualize the characters, you have to hear what their voices sound like." Reflect back on classic and contemporary works which you have read over time, and identify a favorite character that made an impression on you, good or bad. Describe this character in detail and give reasons why he/she was unforgettable.

Standard Two

Historical fiction is a favorite genre for many young adults. Select and read a book that falls into this category. As you enjoy the story, look for objects or symbols that are important to both characters and plot. Symbols are those things that have a special meaning and significance to them. Create a story quilt using those symbols. Draw each symbol and then briefly describe its importance to the story. Paste these on a larger piece of paper as patchworks on a quilt. You may want to fill in spaces between each symbol with plain patches of construction paper.

Standard Three

Talented authors pride themselves on their ability to write all kinds of interesting sentences to tell their story. These sentences vary in purpose, length, and use of colorful words and images. Locate and copy down a sentence that falls into each of these categories from a book or short story that you are currently reading or have just completed: (1) Sentence that relates to one of the five senses of sight, smell, touch, taste, or hearing; (2) Sentence that shows a sense of humor; (3) Sentence with a simile, metaphor, or other figure of speech; (4) Sentence of detailed description; (5) Sentence with interesting dialogue.

Standard Four

Create a series of different answering machine or voice mail messages that you might use for different types of callers. What message would you leave on your cell or home telephone for each of these individuals: Your parents; Your best friends; Your coach or teacher; Your worst enemy; Your unwanted solicitors; Your classmates or acquaintances who have limited English. How does each message vary with the audience for which it was intended?

Standard Five

Compose a persuasive paragraph to answer this question: Who has it easier in the world today, girls or boys, and why?

Standard Six

Write about something interesting that happened to you recently in school or at home. Make certain that your incident is written with many spelling and grammar errors in its composition. Then, rewrite your work with all corrections made. Finally, share your piece of writing with a peer and see if he/she can find and correct all of the errors to your satisfaction.

Standard Seven

List the continents of the earth and collect information about each one of them. Then, make up a conversation that might take place between any three of these continents. Determine what they might say about their geographic locations, their weather, their natural resources, their wildlife, their inhabitants, and their lifestyles.

Standard Eight

Think about all of the ways that you use the computer and the Internet on any given day or week at home and at school. What are the technological and informational benefits of these tools to you and what are some of the not-so-good things about your dependence on these resources at your disposal?

Standard Nine

What advice would you give to a new student who is moving to your school from another part of the world and who has limited ability to speak your language? What hints or help might you provide this person to make his/her adjustment period a positive experience? Why?

Standard Ten

Explain what you think it would be like to move to another country where you couldn't speak the language or understand much of anything that was being said.

Standard Eleven

Look up the word "protest" in the dictionary and write down its key definitions. Use the word in a sentence. Think of something you would like to protest in your life and write an editorial to summarize your position.

Standard Twelve

If you could start a fan club for someone you admire, who would it be for—an athlete, an entertainer, a politician, an author, a family member? Organize your own fan club and create a newsletter celebrating the status of your hero!

Ignite Student Intellect and Imagination in Language Arts, by Sandra L. Schurr and Kathy L. LaMorte. Published by National Middle School Association, 2007.

7. Book Report Outline for Use With Nonfiction or Reference Book

Knowledge Level

Record the answers to these questions: (1) What is the title of the book? (2) Who wrote, edited, or compiled the information in the book? (3) When was the book published? (4) Where did you find the book?

Comprehension Level

Summarize the main ideas, facts, sections, or subject/topics found and discussed in the book.

Application Level

Select 25 to 50 key words or terms and their definitions from the book. Classify them in some meaningful way.

Analysis Level

Compare your book with another book on the same subject. How are the books alike and how are they different?

Synthesis Level

Suppose that you were to write a new book on this or a related topic. Create an original book jacket and table of contents for this masterpiece!

Evaluation Level

Would you recommend the book to anyone else? Give three to five reasons for your answer.

Ignite Student Intellect and Imagination in Language Arts, by Sandra L. Schurr and Kathy L. LaMorte. Published by National Middle School Association, 2007.

National Middle School Association

National Middle School Association, established in 1973, is the voice for professionals and others interested in the education and well-being of young adolescents. The association has grown rapidly and enrolls members in all 50 states, the Canadian provinces, and 42 other nations. In addition, 57 state, regional, and provincial middle school associations are official affiliates of NMSA.

NMSA is the only national association dedicated exclusively to the education, development, and growth of young adolescents. Membership is open to all. While middle level teachers and administrators make up the bulk of the membership, central office personnel, college and university faculty, state department officials, other professionals, parents, and lay citizens are members and active in supporting our single mission—improving the educational experiences of 10- to 15-year-olds. This open and diverse membership is a particular strength of NMSA's. The association publishes *Middle School Journal,* the movement's premier professional journal; *Research in Middle Level Education Online; Middle Ground, the Magazine of Middle Level Education; The Family Connection*, an online newsletter for families; *Classroom Connections,* a practical quarterly resource; and a series of research summaries.

A leading publisher of professional books and other resources in the field of middle level education, NMSA provides materials both for understanding and advancing various aspects of the middle school concept and for assisting classroom teachers in planning for instruction and for assisting principals in providing leadership. More than 70 NMSA publications are available through the resource catalog as well as selected titles published by other organizations.

The association's highly acclaimed annual conference has drawn many thousands of registrants every fall. NMSA also sponsors other professional development opportunities.

For information about NMSA and its many services, contact the association's headquarters office at 4151 Executive Parkway, Suite 300, Westerville, Ohio 43081. TELEPHONE: 800-528-NMSA; FAX: 614-895-4750; INTERNET: www. nmsa.org.